BRIEF REVIEW in

Earth Science

ORDER INFORMATION

Send orders to:

PRENTICE HALL SCHOOL DIVISION
CUSTOMER SERVICE CENTER
4350 Equity Drive
P.O. Box 2649
Columbus, OH 43272

or

CALL TOLL-FREE: 1-800-848-9500
(8:00 AM–4:30 PM)

- Orders processed with your call. Please include the ISBN number on the back cover when ordering.
- **Your price includes all shipping and handling.**

PRENTICE HALL Textbook Programs that help you meet the requirements of the Regents:

Biology: The Study of Life
Biology by Miller and Levine
Chemistry: The Study of Matter
Chemistry: Connections to Our Changing World
Physics: Its Methods and Meanings
Earth Science: A Study of a Changing Planet

BRIEF REVIEW in
Earth Science

JEFFREY C. CALLISTER
Newburgh Free Academy
Newburgh, New York

Prentice Hall

Needham, Massachusetts   Upper Saddle River, New Jersey

ISBN 0-13-436111-3

1 2 3 4 5 6 7 8 9 10   02 01 00 99 98

CONTENTS

ABOUT THIS BOOK

Brief Review in Earth Science is a concise text and review aid for the current New York State Syllabus in Earth Science and a means of preparing for the Regents Examination in this subject. The following features will be of special interest to teachers and students.

1. Brief, simple explanations of all concepts and major understandings included in the syllabus.

2. Organization by *Topics* closely following the syllabus.

3. Particular emphasis on essential vocabulary. Each important term is printed in **bold type** where it is defined and discussed in the text. These terms are also presented for review at the end of each Topic and are defined in the *Glossary*.

4. An abundance of illustrations. There are more than 100 drawings in the text alone, each designed to help students visualize and understand the concepts and vocabulary of the subject. They also familiarize students with the types of diagrams they will be required to interpret in Regents Examination questions.

5. Hundreds of practice questions covering all the major understandings of the syllabus. Some questions are reproduced from past Regents Examinations, while other questions are of Regents type. The questions have been selected to review specific concepts and are grouped in the order in which concepts are discussed within each Topic. Many questions review the newest changes in the *Earth Science Reference Tables.*

6. Four complete Regents Examinations. The four most recent examinations are included in their entirety. For convenience in using these exams as an additional source of questions for each Topic, a question number key is provided at the beginning of the Questions section of each Topic.

7. An exceptionally complete index. The index has been carefully planned and compiled to make it easy for the student to locate the portion of the text where any specific term is defined and any specific subject matter is treated.

I believe this book will be highly effective in helping every student achieve mastery of the course content and objectives of the New York State Syllabus in Earth Science. The author would appreciate suggestions on how to make future editions of this book even more useful.

I would like to acknowledge the invaluable contributions of Vernon G. Abel, Bertram Coren, Lois B. Arnold, and my wife Angie to earlier editions of this book. I would also like to thank Tim Denman, Elizabeth A. Jordan, Ronald Macey, and hundreds of students for their suggestions regarding this newest edition.

Jeffrey C. Callister

TOPIC I Observation and Measurement

OBSERVATION. An **observation** occurs through the interaction of one or more of the **senses**—sight, hearing, touch, taste, or smell—with a part of the environment. The ability of the senses to make observations is limited in range and precision. **Instruments** are devices invented by people to extend the senses beyond their normal limits and thus enable them to make observations that would otherwise be impossible or highly inaccurate. For example, a microscope makes it possible to see objects and details that are too small for the unaided eye to detect; a magnet allows one to observe something that the senses do not respond to at all.

INFERENCE. An **inference** is an interpretation of observations. It is a mental process that proposes causes or explanations for what has actually been observed. For example, the observation of an impression in mud, shaped like a dog's foot, leads to the inference that a dog has been present. This inference may or may not be correct. Additional observations may make the inference more likely to be true. A prediction of an event, such as an earthquake or snowstorm, is also a type of inference.

CLASSIFICATION. Scientists group similar things together in order to make the study of objects and events in the environment more meaningful. This grouping is called **classification,** and it is based on the observed properties of the objects or events.

MEASUREMENT. A **measurement** is a means of expressing an observation with greater accuracy or precision. It provides a numerical value for some aspect of the object or event being observed. Every measurement includes at least one of the three basic *dimensional quantities*—length, mass, and time.

Length may be defined as the distance between two points.

Mass is the quantity of matter in an object. It is often determined by weighing the object, but mass should not be confused with *weight,* which is the pull (force) of the earth's gravitation on an object. The weight of an object may vary with its location, but its mass remains the same.

Time may be described as our sense of things happening one after another or as the duration of an event.

Some types of measurements require mathematical combination of basic dimensional quantities. For example, a unit of volume is actually a unit of length cubed, as in 25 cm^3 (25 cubic centimeters). Other examples are density (mass per unit volume, as in 4 g/cm^3); pressure (force per unit area—15 lb/ft^2); speed (distance per unit time—9.8 km/sec).

All measurements of the basic quantities (length, mass, and time) are made by a direct comparison with certain accepted *standard units* of measurement. For example, length is measured by comparing it with a standard unit such as the centimeter, mass with a unit such as the gram, and time with a unit such as the second. A measurement must always state the units used; for example, 27.9 *grams.*

ERROR.   No measurement is perfect, because it is limited by the imperfection of the senses and of instruments. Human error may also result from carelessness or from improper use of an instrument. Any measurement is therefore an approximation of a true, or absolute, value and must be considered to contain some error.

PERCENT DEVIATION.   Often in science there is an accepted value for a given quality (for example, the density of water). It is then possible to determine the accuracy, or amount of **error,** of a given measurement by comparing it with the accepted value. When the amount of error is expressed as a percentage, it is called **percent deviation** (or **percent error.**) The percent deviation is obtained by dividing the difference between the measured and accepted values by the accepted value and multiplying the result by 100%.

$$\text{percent deviation} = \frac{\text{difference between measured value and accepted value}}{\text{accepted value}} \times 100\%$$

For example, suppose a student measures the mass of an object as 127.5 grams and the accepted value is 125.0 grams. Then,

$$\begin{aligned}
\text{amount of deviation} &= \text{difference between measured value and accepted value} \\
&= 127.5 \text{ grams} - 125.0 \text{ grams} \\
&= 2.5 \text{ grams} \\
\text{percent deviation} &= \frac{\text{amount of deviation}}{\text{accepted value}} \times 100\% \\
&= \frac{2.5 \text{ grams}}{125.0 \text{ grams}} \times 100\% \\
&= 2.0\%
\end{aligned}$$

DENSITY.   The concentration of matter in an object is known as its **density.** The density of an object is the ratio of its mass (quantity of matter) to its volume; that is, density is the mass in each unit of volume. (**Volume** is the amount of space an object occupies.) To determine the density of an object, its mass is divided by its volume:

$$\text{density} = \frac{\text{mass}}{\text{volume}}$$

The density of a material does not depend on the size or shape of the sample, as long as the temperature and pressure remain the same. For example, a cube of aluminum with a volume of 20 cm^3 has a mass of 54 grams. Its density is 54 g/20 cm^3 = 2.7 g/cm^3. An aluminum ball with a volume of 40 cm^3 (twice the volume of the cube) will have a mass of 108 g (twice the mass of the cube), but its density will be the same: $108 \text{ g}/40 \text{ cm}^3 = 2.7 \text{ g/cm}^3$.

FLOTATION.   An object immersed in a liquid is buoyed up by a force equal to the weight of liquid it displaces (Archimedes' Principle). If the density of the object is less than the density of the liquid, the weight of displaced liquid will be greater than the weight of the object. Therefore, the upward force will be enough to support the object, and it will float in the liquid. Flotation of objects in liquids (and in gases) is one method of determining their relative densities. The lower the density of a floating object, the higher it floats in the liquid, that is, the greater the percentage of its volume that is above the surface. As the density of a floating object increases, it sinks relatively deeper into the liquid. If its density is greater than that of the liquid, it will sink to the bottom. If an object and a liquid have exactly the same density (for example, a fish in water), the object can remain stationary anywhere in the liquid.

FACTORS THAT AFFECT DENSITY.   Changes in temperature and pressure affect the density of substances, especially of gases. If the temperature of a gas increases, and its pressure remains the same, its molecules move farther apart (that is, the gas expands). The temperature rise thus results in less mass per unit volume, so that the density of the gas decreases. This explains why hot air rises when surrounded by cooler air.

If the pressure on a gas increases, the molecules come closer together (the gas contracts). The pressure rise thus results in more mass per unit volume, and the density increases.

The expansion and contraction caused by temperature and pressure variations occur also in solids and liquids, but to a much smaller extent. Therefore the density of solids and liquids is less affected by such environmental changes.

PHASES OF MATTER.   Matter on earth exists in three main forms: solid, liquid, and gaseous. Each form is known as a **phase,** or **state,** of matter. The density of a substance changes with changes in its phase. Most substances increase in density as they change from a gas to a liquid, and from a liquid to a solid. They have their highest density as a solid because the atoms are closest together in that phase. Water is an exception, having its highest density in the liquid state at a temperature of 4° C. Therefore, solid water (ice) floats on liquid water, while in most substances the solid sinks in the liquid. Water at 4° C will also lie below layers of water at any other temperature.

VOCABULARY

observation	measurement	density
senses	mass	volume
instrument	error	phase (state)
inference	percent deviation or	
classification	percent error	

QUESTIONS ON TOPIC I—
OBSERVATION AND MEASUREMENT

Questions in Recent Regents Exams (end of book)

June 1995: 1, 2, 3, 102, 103
June 1996: 2, 3, 4, 20
June 1997: 3, 56–60
June 1998: 1, 57, 59, 60, 66, 85

Questions from Earlier Regents Exams

1. Using a ruler to measure the length of a stick is an example of (1) extending the sense of sight by using an instrument   (2) calculating the percent of error by using a proportion   (3) measuring the rate of change of the stick by making inferences   (4) predicting the length of the stick by guessing

2. Which statement made by a student after examining a rock specimen is an inference?   (1) The rock is of igneous origin.   (2) The rock has rounded edges.   (3) The rock is light-colored.   (4) The rock contains large crystals.

3. Which descriptive term illustrates an inference?   (1) transparent (2) bitter   (3) younger   (4) smooth

4. A prediction of next winter's weather is an example of   (1) a measurement   (2) a classification   (3) an observation   (4) an inference

5. A classification system is based on the use of   (1) the human senses to observe properties of objects   (2) instruments to observe properties of objects   (3) observed properties to group objects with similar characteristics   (4) inferences to make observations

6. A measurement is best defined as   (1) an inference made by using the human senses   (2) a direct comparison with a known standard (3) an interpretation based on theory   (4) a group of inferred properties

7. Using the centimeter ruler provided in the *Earth Science Reference Tables*, what is the length of line segment *X-Y*?

(1) 5.8 cm   (2) 6.1 cm   (3) 6.4 cm   (4) 7.1 cm

8. The diagram below represents the scale of a triple beam balance. What mass is indicated by this scale?

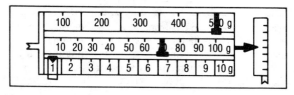

(1) 130 g   (2) 151 g   (3) 517 g   (4) 571 g

9. A student determines the density of an ice cube to be 0.80 gram per cubic centimeter, when it is actually 0.90 gram per cubic centimeter. What is the percent deviation (percent error) in this calculation? [Refer to the *Earth Science Reference Tables.*]   (1) 6%   (2) 11%   (3) 13%   (4) 88%

10. Which term is best defined as a measure of the amount of space a substance occupies?   (1) mass   (2) volume   (3) density   (4) weight

11. Which is the best definition for the mass of an object?   (1) the amount of space the object occupies   (2) its ratio of weight to volume   (3) the quantity of matter the object contains   (4) the force of gravity acting on the object

12. Which characteristic of an object will always change as the object travels from the earth to the moon?   (1) mass   (2) volume   (3) density   (4) weight

Base your answers to questions 13 through 17 on your knowledge of earth science, the *Earth Science Reference Tables*, and the diagrams below. Objects *A* and *B* are solid and made of the same uniform material.

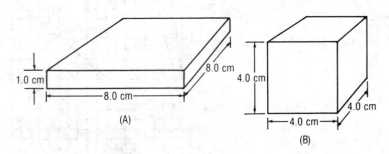

13. If object *B* has a mass of 173 grams, what is its density? (1) 0.37 g/cm^3 (2) 2.7 g/cm^3 (3) 3.7 g/cm^3 (4) 5.7 g/cm^3

14. Object *A* expands when it is heated. Which graph best represents the relationship between the temperature and the density of object *A*?

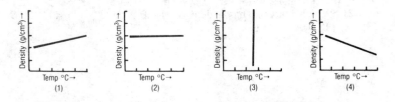

15. A student measures the mass of object *B* to be 156 grams, but the actual mass is 173 grams. The student's percent deviation (percentage of error) is approximately   (1) 1%   (2) 5%   (3) 10%   (4) 20%

16. A third object is made of the same uniform material as object B, but has twice the volume as object B. How does the density of this third object compare to the density of object B?   (1) It is one-half as dense as B.   (2) It is the same density as B.   (3) It is twice as dense as B. (4) It is four times as dense as B.

Note that question 17 has only three choices.

17. How does the mass of object B compare to the mass of object A?   (1) The mass of B is less than the mass of A.   (2) The mass of B is greater than the mass of A.   (3) The mass of B is the same as the mass of A

Base your answers to questions 18 through 21 on your knowledge of earth science and on the graph at the right, which shows the masses and volumes of four different earth materials.

18. Which material has the greatest density?
(1) A
(2) B
(3) C
(4) D

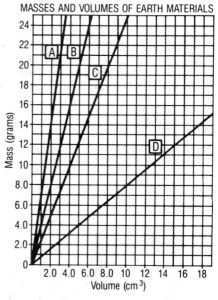

MASSES AND VOLUMES OF EARTH MATERIALS

19. If the density of water is 1 g/cm³, which material will float on water?   (1) A (2) B (3) C (4) D

20. What is the mass of sample C if its volume is 3.0 cubic centimeters?   (1) 7.5 g   (2) 2.1 g   (3) 3.0 g   (4) 21 g

21. Which material has a density of 4.0 g/cm³?   (1) A (2) B (3) C (4) D

22. The two solid blocks below are made of the same material and are under the same temperature and pressure conditions. If the mass of block X is 54 grams, what is the mass of block Y?

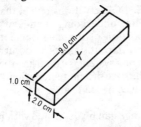 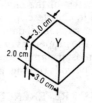

(1) 18 g   (2) 27 g   (3) 54 g   (4) 108 g

23. Substances A, B, C, and D are at rest in a container of liquid as shown by the diagram below. Which choice lists the substances in the order of lowest density to highest density?

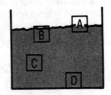

(1) A, B, C, D (2) A, D, C, B (3) D, C, B, A (4) C, B, A, D

24. Substances A, B, C, and D are at rest in a container of liquid as shown in the accompanying diagram. Which substance probably has the same density as the liquid?

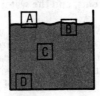

(1) A (2) B (3) C (4) D

25. In which phase (state) do most earth materials have their greatest density?   (1) solid   (2) liquid   (3) gaseous

26. Which graph represents the behavior of water as it is warmed from 0° C to 16° C?

CHARACTERISTICS OF CHANGE.   Observations reveal that most of the properties of the environment are undergoing **change** at all times. As a result, the description of a portion of the environment is never exactly the same in all details at two different times. The occurrence of a change in the properties of an object or a system is called an **event.** Events may be almost instantaneous, as in the case of a meteor, or they may occur over long periods of time, as in changes in sea level.

FRAMES OF REFERENCES.   Change can be described with respect to time and space (location). Time and space are called the **frames of reference** for studying change. As an example, we say that the moon changes because we observe it in different locations and different phases at different times.

RATE OF CHANGE.   **Rate of change** describes how much a measurable aspect of the environment (a field) changes in a period of time (years, hours, or seconds). Rate of change is determined by use of the following equation, which is found in the *Earth Science Reference Tables.*

$$\text{rate of change} = \frac{\text{change in field value}}{\text{change in time}}$$

CYCLIC CHANGE.   Many changes in the environment occur in some orderly fashion in which the events constantly repeat themselves with reference to space and time. Such orderly changes are called **cyclic change,** and they include the movement of celestial objects (sun, moon, stars, planets), seasonal events, and the water and rock cycles.

PREDICTION OF CHANGE.   A prediction is a type of inference about the conditions of the environment in the future. Predictions can be made if the amount, type, and direction of change can be determined. Because eclipses are cyclic, for example, astronomers can predict the occurrence of solar and lunar eclipses many hundreds of years in the future.

ENERGY AND CHANGE.   All change involves a flow of energy from one part of the environment (which loses energy) to another part (which gains it). For example, in the movement of the water in a stream, friction causes the stream to lose energy, while the environment around the stream gains energy (in the form of heat). Energy is usually exchanged across an **interface,** which is the boundary between regions with different properties. In the case of the stream, there is an interface between the water and the air and another between the water and the stream bed.

ENVIRONMENTAL EQUILIBRIUM. Although change occurs continuously throughout the environment, certain general characteristics tend to remain constant. For example, the wooded shore of a lake tends to look the same from one day to the next. Although it may go through a cycle of seasonal changes, it tends to look the same each year at the same season. This is the result of a natural balance among all the changes taking place, called **environmental equilibrium.** This equilibrium is easily upset on a small scale, for example by the burrowing of a worm in the soil, but is normally not upset on a large scale. Human activities, however, often abruptly upset the natural environmental equilibrium. For example, construction equipment can rapidly change a rain forest into a city, thus introducing change on a large scale.

POLLUTION OF THE ENVIRONMENT. **Pollution** of the environment occurs when the concentration of any substance or form of energy reaches a proportion that adversely affects people, their property, or plant or animal life.

CAUSES OF POLLUTION. Many forms of pollution are the result of human technology, which often produces substances and forms of energy in harmful quantities. Some forms of pollution are the results of natural events or processes and would occur without the presence of people. High concentrations of pollen in the air, and ash and gases from volcanic eruptions, are examples of natural pollution.

Pollutants (things that pollute) include solids, liquids, gases, biologic organisms, and forms of energy such as heat, sound, and nuclear radiation. People, through their technology, add pollutants by individual action, community action, and by industrial processes. Many forms of pollution vary with the time of day or season of the year because the causes are cyclic. Air pollution in cities, for example, may increase at the times people are going to and from work because the automobile is one of the greatest contributors to air pollution.

VOCABULARY	
change	interface
event	environmental equilibrium
rate of change	pollution
frames of reference	pollutants
cyclic change	

QUESTIONS ON TOPIC II—THE CHANGING ENVIRONMENT

Questions in Recent Regents Exams (end of book)

Questions from Earlier Regents Exams

1. What always happens when a change occurs?   (1) Pollution is produced.   (2) The temperature of a system increases.   (3) The properties of a system are altered.   (4) Dynamic equilibrium is reached.

2. Which statement best explains why some cyclic earth changes may *not* appear to be cyclic?   (1) Most earth changes are caused by human activities.   (2) Most earth changes are caused by the occurrence of a major catastrophe.   (3) Many earth changes occur over such a long period of time that they are difficult to measure.   (4) No earth changes can be observed because the earth is always in equilibrium.

3. Which graph represents the greatest rate of temperature change?

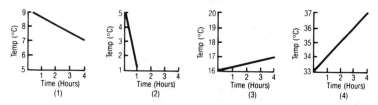

4. Refer to the data table below. At approximately what rate did the temperature rise inside the greenhouse between 8:00 A.M. and 10:00 A.M.?

Time	Average Inside Temperature
6:00 A.M.	13°C
8:00 A.M.	14°C
10:00 A.M.	16°C
12:00 noon	20°C

(1) 1.0°C/hour   (2) 2.0°C/hour   (3) 0.5°C/hour   (4) 12.0°C/hour

5. The rising and setting of the sun are examples of   (1) noncyclic events   (2) unrelated events   (3) predictable changes   (4) random motion

6. Future changes in the environment can best be predicted from data that are   (1) highly variable and collected over short periods of time   (2) highly variable and collected over long periods of time   (3) cyclic and collected over short periods of time   (4) cyclic and collected over long periods of time

7. An interface can best be described as   (1) a zone of contact between different substances across which energy is exchanged   (2) a region in the environment with unchanging properties   (3) a process that results in changes in the environment   (4) a region beneath the surface of the earth where change is not occurring

8. Which condition exists when the rates of water flowing into and out of a lake are balanced so that the lake's depth appears to be constant? (1) equilibrium   (2) transpiration   (3) hydration   (4) saturation

9. Why were laws passed that made some insecticides, such as DDT, unavailable to the general public?   (1) Some insecticides were not effective against insects.   (2) Some insecticides' colors were causing a change in the environment's physical appearance.   (3) Some insecticides' concentration in the environment became harmful to other animal life as well as to insects.   (4) Some insecticides' effects were weakened when diluted by the atmosphere and hydrosphere.

10. When the amounts of biologic organisms, sound, and radiation added to the environment reach a level that harms people, these factors are referred to as environmental   (1) interfaces   (2) pollutants (3) phase changes   (4) equilibrium exchanges

11. People who live close to major airports are most likely to complain about which form of pollution?   (1) sound   (2) heat   (3) radioactivity   (4) particulates

12. Which energy source is *least* likely to pollute the environment? (1) uranium   (2) petroleum   (3) coal   (4) wind

Additional Questions

1. Which statement is *not* true about earth changes?   (1) Most changes are cyclic.   (2) The scope and direction of most changes are not predictable.   (3) Change is the natural state of the environment. (4) A change in the environment also involves a change in time and space quantities.

2. Most changes in the environment are   (1) cyclic   (2) destructive (3) sudden   (4) unnatural

3. Which statement is true about the environment of most of the earth?   (1) It is usually greatly unbalanced.   (2) It is usually in equilibrium and hard to change.   (3) It is usually in equilibrium but easy to change on a small scale.   (4) It is normally heavily polluted by natural occurrences.

4. Why have humans been such a big factor in the pollution of the environment?   (1) They have had many wars.   (2) They are larger in size than most animal forms.   (3) Their technology creates pollutants. (4) They are the only form of life that can adapt to almost all the environments on the earth's surface.

5. Which statement is true about the pollen, carbon monoxide, and sound found in an urban community?   (1) Their concentrations change with time.   (2) They are not forms of pollution because they can be caused by natural events.   (3) They are always considered pollutants because they adversely affect the environment in any concentration. (4) They are produced by the technology of people and are thus considered forms of pollution.

TOPIC III   Measuring the Earth

SIZE AND SHAPE OF THE EARTH

THE EARTH'S SHAPE. The earth's shape is an **oblate spheroid;** that is, it is a sphere with a slight flattening at the polar regions and a slight bulging at the equatorial region. As a result, the equatorial diameter and circumference are larger than the polar diameter and circumference respectively, as Figure 3-1 shows.

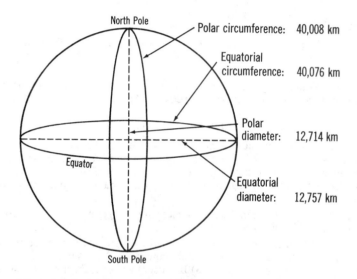

Figure 3-1. Polar and equatorial dimensions of the earth.

However, the differences are very small compared to the actual dimensions; for example, the difference between the two circumferences is only 68 kilometers. This variation from a spherical shape could not be detected in a model of the earth, such as an ordinary globe. Such models are therefore made in the shape of a sphere. (A **model** is any way of representing the properties of an object or system. A model may be an object, such as a globe to represent the earth, or it may be a drawing, diagram, graph, chart, table, or even a mathematical formula or equation. Scientists construct models of these various kinds as aids in studying and understanding the environment.) Besides being very close to a perfect sphere, the earth's surface is very smooth (has very little relief) compared to its diameter.

EVIDENCE FOR THE EARTH'S SHAPE

1. Altitude of Polaris. The **altitude** of an object in the sky is its angle above the horizon. The **latitude** of a point on the earth's surface is its angle north or south of the equator. The star **Polaris** (also called the *North Star*) is almost directly over the North Pole of the earth. From the geometry of a sphere, it can be shown that the altitude of Polaris at any point in the Northern Hemisphere should be the same as the latitude of that point. That is, the altitude of Polaris should change from 0° at the equator to 90° at the North Pole. In the Northern Hemisphere, the altitude of Polaris is equal to the latitude of the observer. (See Figure 3-2).

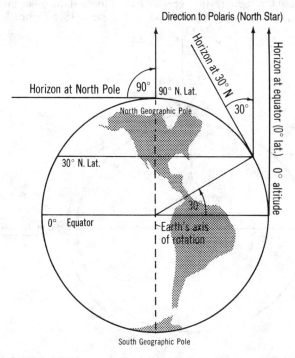

Figure 3-2. Showing that the altitude of Polaris should be the same as the latitude of the observer on a spherical earth. Note that the **geographic poles** are at the ends of the earth's **axis** of rotation.

The altitude of Polaris, however, does not vary exactly in step with latitude. The observed variation is evidence that the earth is not a perfect sphere but is slightly oblate.

2. Photographs from space. Extremely precise photographic observations of the earth taken from satellites and manned spacecraft confirm the oblateness of the earth's shape.

3. Gravity measurements. *Gravity* is the force that attracts all objects toward the center of the earth. The amount of this force on an object is called the object's *weight*. The law of gravitation states that this force is inversely proportional to the square of the distance between the centers of the earth and the object. That is, as the distance from the center of the earth increases, the weight of an object decreases. On a perfectly spherical earth, the weight of a body should be the same at any latitude, since all points of the surface of a sphere are the same distance from the center. However, measurements show that an object weighs more near the poles than near the equator. Part of the reason for this is an apparent outward force caused by the earth's rotation. But when this is accounted for, there is still a greater gravity force at the poles than at the equator, indicating an oblate shape for the earth. (See Figure 3-3).

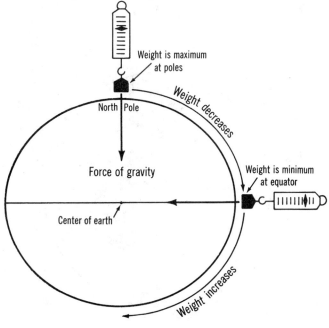

Figure 3-3. Gravity varies with latitude. Because of the flattening of the earth at the poles the distance from the center of the earth is less at the poles than at the equator. As a result, the force of gravity is greater at the poles than at the equator. (Flattening greatly exaggerated in this drawing.)

4. "Sinking" ships. As a ship sails away from shore, it appears to sink at the horizon. Although this effect does not prove that the earth is spherical, it is evidence that the earth's surface is curved.

MEASURING THE EARTH. In our space age, measurements of the earth's dimensions can be made from space. However, it has long been possible to estimate the earth's circumference quite accurately from earth by using fairly simple methods.

About 200 B.C. a Greek scientist named Eratosthenes measured the circumference of the earth using angles of the sun at noon. (Refer to Figure 3-4.) Eratosthenes knew that on a particular day each year the sun shone directly down to the bottom of a well in one city in Egypt. In another city due north, the sun cast a shadow at noon on that day. Eratosthenes measured the distance between the two cities and calculated the earth's circumference using the following equation, which is also found in the *Earth Science Reference Tables.*

$$\frac{\text{shadow angle}}{360°} = \frac{\text{distance of surface}}{\text{circumference}}$$

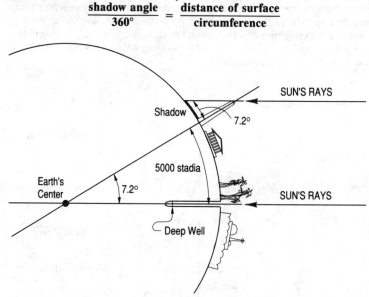

Figure 3-4. Determining the circumference of the earth.

Using this formula, Eratosthenes calculated a circumference equal to about 40,000 km, which is close to present-day measurements. Once the circumference of the earth is estimated, its radius, diameter, surface area, and volume can also be calculated (see Figure 3-5).

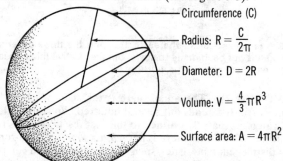

Circumference (C)

Radius: $R = \dfrac{C}{2\pi}$

Diameter: $D = 2R$

Volume: $V = \dfrac{4}{3}\pi R^3$

Surface area: $A = 4\pi R^2$

Figure 3-5. Relationships among the dimensions of a sphere.

THE OUTER PARTS OF THE EARTH

THE LITHOSPHERE.   The **lithosphere** is a layer of rock that forms the solid outer shell of the earth (see Figure 3-6). The lithosphere is approximately 100 km thick. The upper portion is called the crust.

THE HYDROSPHERE.   The **hydrosphere** is the layer of water that rests on the lithosphere. The hydrosphere consists of the ocean, which covers about 70% of the earth's surface, plus the water in bodies such as lakes, streams, and rivers. The ocean layer is relatively thin, averaging only 3.8 kilometers in thickness; in fact, even the thin line labeled "hydrosphere" in Figure 3-6 is too thick to show the hydrosphere in correct relation to the lithosphere and atmosphere.

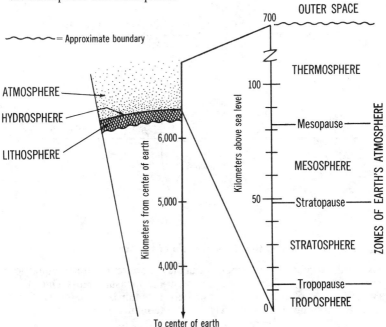

Figure 3-6. A cross section of the earth showing the three outer spheres. The drawing has not been made to scale, but the relative thicknesses of the parts of the earth are close to proportion.

THE ATMOSPHERE   The **atmosphere** is the shell of gases that surrounds the earth. It extends out several hundred kilometers into space, but nearly all of its mass is confined to the first few kilometers from the earth's surface. The atmosphere is *layered,* or *stratified,* into zones with their own distinct characteristics, such as temperature and composition. (See *Selected Properties of Earth's Atmosphere* in the *Earth Science Reference Tables.*)

LOCATING POSITIONS ON THE EARTH

COORDINATE SYSTEMS. To fix the location of a point on any two-dimensional surface, such as the surface of the earth, two numbers, called *coordinates,* are needed. The system for determining the coordinates of a point is called a **coordinate system.** The *latitude-longitude system* is the one commonly used to locate points on the earth. Latitude and longitude are measured in angular units: degrees (°), minutes ('), and seconds (").

LATITUDE. Latitude is an angular distance north or south of the equator. If a line is drawn from any point on the earth's surface to the center of the earth, the latitude of that point is the number of degrees in the angle between the line and the plane of the equator. All points that have the same latitude lie on a circle that is parallel to the equator. These circles are called **parallels** of latitude. The **equator** may be considered to be the parallel of latitude 0°. Latitude increases north and south of the equator to a maximum of 90° at each pole.

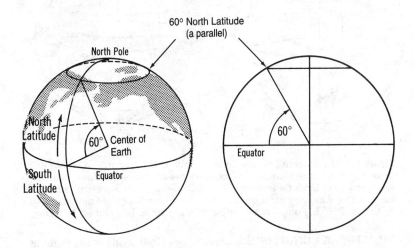

Figure 3-7. The meaning of latitude. Latitude is the angular distance north or south of the equator. On the left, the earth is shown as a sphere with the 60° parallel of North Latitude shown as a circle parallel to the equator. On the right, the earth is shown with the North-South axis vertical; the equator and the parallels of latitude then becomes horizontal lines.

MEASURING LATITUDE. As shown in Figure 3-2 on page 13, the altitude of Polaris equals the degree of latitude. In the Southern Hemisphere, latitude can be determined by measuring the altitude of certain other stars. In both cases, astronomical tables are used to make adjustments for date and time if extreme precision is desired.

LONGITUDE. Longitude is an angular distance east or west of the Prime Meridian. A **meridian** is any semicircle on the earth's surface connecting the North and South Poles. The meridian that passes through Greenwich, England, has been designated the *Prime Meridian,* or the meridian of zero longitude. The longitude of any point on the Prime Meridian is 0°. The longitude of any other point on the earth's surface is the number of degrees between the meridian that passes through the point and the Prime Meridian. This angle can be measured along the equator or along any parallel of latitude. Since a full circle is 360°, longitude increases east or west of the Prime Meridian from 0° at the Prime Meridian to 180° at the meridian that is the continuation of the Prime Meridian on the other side of the earth.

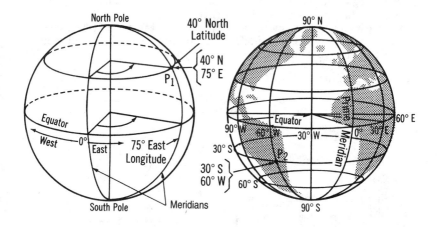

Figure 3-8. The meaning of longitude. Longitude is the angular distance east or west of the Prime Meridian, which is the meridian passing through Greenwich, England. Latitude and longitude together provide a system of coordinates for locating any point on the earth. The coordinates of Point P_1 are 40° N. Lat. and 75° E. Long. Those for Point P_2 are 30° S. Lat. and 60° W. Long.

MEASURING LONGITUDE. *Local noon* (12:00 noon) at any point on the earth occurs when a line from the sun to the center of the earth cuts the meridian of that point. At that moment the sun reaches its highest altitude of the day, so that the instant of local noon can be determined by observing the sun. Since the earth rotates from west to east at the rate of one rotation per day, or 360° in 24 hours, it rotates 15° per hour. Therefore, the occurrence of local noon moves from east to west at the same rate of 15° per hour. Longitude can be calculated if, when local noon occurs, the observer knows what time it is at Greenwich, England. For example, if local noon occurs at 1:00 P.M. Greenwich time (G.M.T.), one hour has passed since the sun crossed the Prime Meridian; the local longitude is therefore 15° west. In general, longitude can be calculated by finding the time difference in hours between local sun time and Greenwich time, and multiplying by 15°. If local time is earlier than Greenwich time, the longitude is west; if later, it is east. Greenwich time can be determined if the observer has a clock that has been set to keep Greenwich time, or by means of radio signals that are broadcast regularly for that purpose.

FIELDS

DEFINITION.   A **field** is any region of space or the environment that has some measurable value of a given quantity at every point. Examples of field quantities are gravity, magnetism, elevation or depth, atmospheric pressure, wind, and relative humidity.

VECTOR AND SCALAR FIELDS.   Some fields need only an amount, or *magnitude,* to be completely described. These fields are called **scalar fields** and include such things as temperature, atmospheric pressure, and relative humidity. Other fields need a magnitude plus a *direction* to totally describe them. These fields are called **vector fields** and include such things as magnetism, gravity, and wind.

ISOLINES.   The varying values of a field are often represented on a map of the region by the use of lines, called **isolines,** which connect points of equal value. Some common examples of isolines are *isotherms* (which connect points of equal temperature), *isobars* (which connect points of equal air pressure), and *contour lines* (which connect points of equal elevation).

ISO-SURFACES.   Isolines show field values over a *two-dimensional* field or surface. For example, isotherms on a weather map show temperatures at one level only (usually ground level). To show field values throughout a *three-dimensional* region or volume of space, it is necessary to use *iso-surfaces.* An **iso-surface** is a surface all of whose points have the same field value. (See Figure 3-9 on page 20.)

CHANGES IN FIELDS.   Since the environment is constantly changing, field characteristics usually change with time. This means that any model of a field, such as a weather map, shows the field for only one particular time.

GRADIENT.   The rate at which a field changes from place to place within the field is called the **gradient** or *slope.* It can be calculated from the following formula:

$$\text{gradient} = \frac{\textbf{change in (amount of) field value}}{\textbf{change in distance}}$$

For example, if a weather map shows a change in temperature of 12°C between two locations 3 km apart, the gradient between the two locations is 4°C per km.

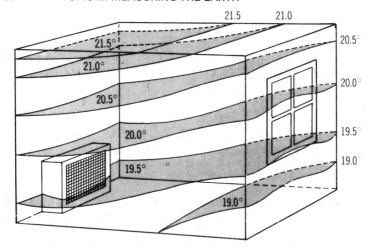

Figure 3-9. Isothermal surfaces in the 3-dimensional temperature field inside a room. All points in any of the iso-surfaces have the same temperature. Where the iso-surfaces meet the ceiling, a 2-dimensional field of isothermal lines is formed. A similar 2-dimensional field can be seen in the right-hand wall.

USING CONTOUR MAPS

CONTOUR MAPS. A **contour map** (topographic map) is a commonly used model of the elevation field of the surface of the earth. The **contour lines** are isolines that connect points of equal elevation above sea level. Other symbols are used to show other features.

The difference in elevation between consecutive contour lines is the *contour interval*. In reading contour lines, pay attention to these points: (1) When contour lines cross a stream, they bend uphill toward higher elevations. Contour lines "point" upstream. (2) *Depression contours* are marked with small lines pointing toward the center of a depression. (Refer to the upper left-hand corner of Figure 3-10.)

DISTANCE ON MAPS. To find the horizontal distance between two places on a map, you compare the distance on the map with the map scale. For example, the length of the road with the number 84 on the map in Figure 3-10 is approximately 4 miles.

MAP DIRECTION. Most maps, including contour maps, usually show directions by indicating north with some type of arrow. The map in Figure 3-10 indicates north by ★, which stands for "geographic north." On most maps, the top faces north. There is usually a north arrow or a compass rose like the one here.

CALLISTER QUADRANGLE

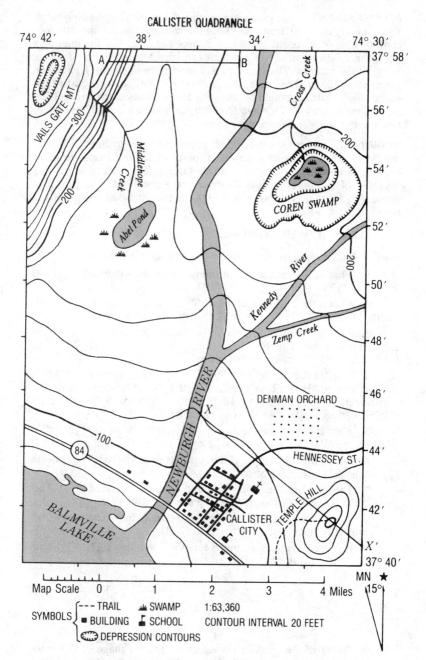

Figure 3-10. A topographic map.

MAP GRADIENT. For practice, compute the gradient of the Newburgh River in Figure 3-10 from the 180-foot contour line to the 100-foot contour line. Your result should be 20 feet per mile. On any map of a field, the relative amount of gradient can be estimated by the distance between the isolines. The more closely spaced the lines, the greater the gradient. In the map example, the gradient is great, or steep, just to the east of Vails Gate Mountain because the contour lines are close together. In the area around the Denman Orchard, the gradient is small, or gentle, because the contour lines are far apart.

PROFILES. The three-dimensional nature and the gradient of a field can be shown by drawing a profile. Profiles of contour maps are often drawn to show the shape of the earth's surface. The method of making a profile of a contour map is illustrated in Figure 3-11. The profile is drawn for the region between X and X' in the southeast corner of the map in Figure 3-10.

TOPOGRAPHIC MAP EXERCISE

Answer the following questions based on the map in Figure 3-10.

    1. What is the maximum altitude of Denman Orchard?   (1) 140 feet   (2) 150 feet   (3) 159 feet   (4) 161 feet

    2. What is the longitude of the middle of Coren Swamp?   (1) 37°53′ North   (2) 37°53′ West   (3) 74°32′ North   (4) 74°32′ West

    3. What is the total distance of Hennessey Street on the map from the margin of the map to where it meets Route 84?   (1) 2½ miles   (2) 3 miles   (3) 3½ miles   (4) 4 miles

    4. What is the gradient of the Kennedy River from the 200-foot contour line to the contour line before the Kennedy River meets the Newburgh River?   (1) 5 to 10 feet/mile   (2) 15 to 20 feet/mile   (3) 25 to 30 feet/mile   (4) 35 to 40 feet/mile

    5. What is the latitude of the school?   (1) 74°34′ North   (2) 74°34′ West   (3) 37°41′ West   (4) 37°41′ North

    6. Which side of Temple Hill has the steepest gradient?   (1) north   (2) south   (3) east   (4) west

    7. The most gentle slope is found in the vicinity of   (1) Abel Pond   (2) Denman Orchard   (3) Temple Hill   (4) Zemp Creek

    8. What direction is Vails Gate Mountain from Callister City?   (1) north   (2) northwest   (3) northeast   (4) west

    9. In what direction is the Kennedy River flowing?   (1) south   (2) southwest   (3) north   (4) northeast

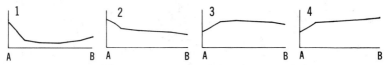

    10. Which of the profiles above represents the shape of the landscape from A to B on the map?   (1) 1   (2) 2   (3) 3   (4) 4

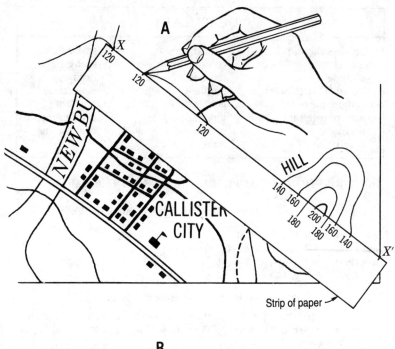

A

X

120

120

120

NEWBU

CALLISTER
CITY

HILL

140 160
180
180 180 160 140

200

X'

Strip of paper

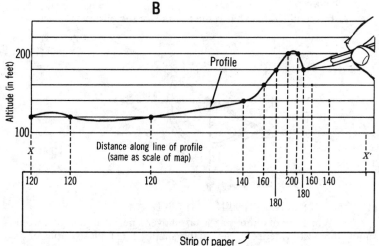

B

Altitude (in feet)

200

Profile

100

X

Distance along line of profile
(same as scale of map)

X'

120   120   120   140   160   200 | 160 140
                                                    180
                                              180

Strip of paper

Figure 3-11. Constructing a profile along the line X-X′ on the contour map of Figure 3-10. The edge of a strip of paper is placed along the line, and a mark is made wherever the paper crosses a contour line. The marks are labeled with the corresponding altitude. The marks are then projected upward to locate dots on a piece of lined paper as in Drawing B. Finally the dots are connected with a smooth curved line, and the profile is complete.

VOCABULARY		
oblate spheroid	atmosphere	scalar field
model	coordinate system	vector field
altitude	latitude	isoline
Polaris (North Star)	parallels	iso-surface
geographic poles	equator	gradient
axis (of rotation)	longitude	contour map
lithosphere	meridians	contour line
hydrosphere	field	

QUESTIONS ON TOPIC III—MEASURING THE EARTH

Questions in Recent Regents Exams (end of book)

Questions from Earlier Regents Exams

1. The true shape of the earth is best described as a   (1) perfect sphere   (2) perfect ellipse   (3) slightly oblate sphere   (4) highly eccentric ellipse

2. In the diagram, what is the latitude of the observer?

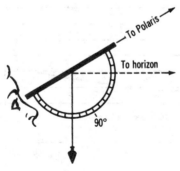

(1) 30°N   (2) 60°N   (3) 90°N   (4) 120°N

3. Which set of photographs taken over a period of one year would supply the best evidence of the earth's shape?   (1) photographs of the sun from the earth   (2) photographs of the earth from the moon   (3) photographs of the earth's shadow on the moon   (4) photographs of the star paths from the North Pole

4. Measurements taken with gravity meters would be most useful in providing information concerning the   (1) magnetic field intensity   (2) shape of the earth   (3) tidal range   (4) distance from the sun

5. Which statement provides the best evidence that the earth has a nearly spherical shape?   (1) The sun has a spherical shape.   (2) The altitude of Polaris changes in a definite pattern as an observer's latitude changes.   (3) Star trails photographed over a period of time show a circular path. ·   (4) The lengths of noontime shadows change throughout the year.

6. The actual polar diameter of the earth is 12,714 kilometers. The equatorial diameter of the earth is approximately   (1) 12,671 km   (2) 12,700 km   (3) 12,714 km   (4)12,757 km

7. Based on the diagram below, what is the circumference of planet Y?

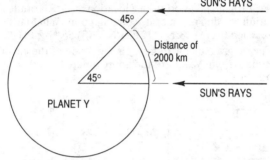

(1) 9,000 km   (2) 12,000 km   (3) 16,000 km   (4) 24,000 km

8. According to the *Earth Science Reference Tables,* the radius of the earth is approximately   (1) 637 km   (2) 6,370 km   (3) 63,700 km   (4) 637,000 km

9. The water sphere of the earth is known as the   (1) atmosphere   (2) troposphere   (3) lithosphere   (4) hydrosphere

10. In the diagrams below, the dark zone at the surface of each wedge-shaped segment of the earth represents average ocean depth. Which segment is drawn most nearly to scale?

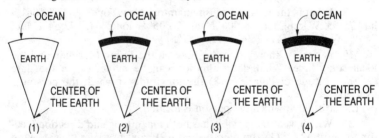

11. What is the approximate altitude of the mesopause in the atmosphere? (1) 50 km (2) 66 km (3) 82 km (4) 90 km

12. Which statement most accurately describes the earth's atmosphere?   (1) The atmosphere is layered, with each layer possessing distinct characteristics.   (2) The atmosphere is a shell of gases surrounding most of the earth.   (3) The atmosphere's altitude is less than the depth of the ocean.   (4) The atmosphere is more dense than the hydrosphere but less dense than the lithosphere.

13. In which group are the spheres of the earth listed in order of increasing density?   (1) atmosphere, hydrosphere, lithosphere   (2) hydrosphere, lithosphere, atmosphere   (3) lithosphere, hydrosphere, atmosphere   (4) lithosphere, atmosphere, hydrosphere

14. As a person travels due west, the altitude of Polaris will (1) decrease   (2) increase   (3) remain the same

15. Polaris is used as a celestial reference point for the earth's latitude system, because Polaris   (1) always rises at sunset and sets at sunrise   (2) is located over the earth's axis of rotation   (3) can be seen from any place on earth   (4) is a very bright star

16. An airplane takes off from a location at 17°S latitude and flies to a new location 55° due north of its starting point. What latitude has the airplane reached?   (1) 28°N   (2) 38°N   (3) 55°N   (4) 72°N

17. The diagram below represents a portion of the earth's latitude and longitude system. What are the approximate latitude and longitude of point A?

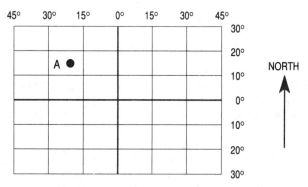

(1) 15°S 20°W   (2) 15°S 20°E   (3) 15°N 20°W   (4) 15°N 20°E

18. What is the location of Binghamton, New York?   (1) 42°06' N lat., 75°55' W long.   (2) 42°06' N lat., 76°05' W long.   (3) 42°54' N lat., 76°05' W long.   (4) 42°54' N lat., 75°55' W long.

19. A person knows the solar time on the Prime Meridian and the local solar time. What determination can be made?   (1) the date   (2) the altitude of Polaris   (3) the longitude at which the person is located   (4) the latitude at which the person is located

20. Which weather variable has both magnitude and direction (vector quantity)?   (1) temperature   (2) air pressure   (3) wind   (4) relative humidity

21. If no elevation values were given, which general rule could be used to establish that a river shown on a contour map flows into a lake? (1) Rivers shown on maps generally flow southward.   (2) Rivers always flow toward large bodies of water.   (3) Contour lines bend upstream when crossing a river.   (4) A large body of water is generally the source of water for a river.

22. The map below represents an elevation field.

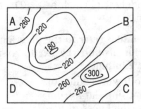

Which graph best represents the elevation profile along a straight line from point *A* to point *B*?

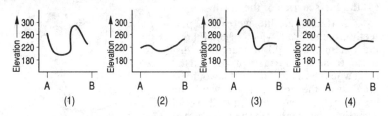

Base your answer to question 23 on the topographic map shown below.

23. What is the most likely elevation of point *A*?   (1) 1,250 (2) 1,650   (3) 1,750   (4) 1,850

Base your answers to question 24 and 25 on the diagram below, which represents a contour map of a hill.

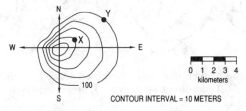

24. On which side of the hill does the land have the steepest slope? (1) north   (2) south   (3) east   (4) west

25. What is the approximate gradient of the hill between points *X* and *Y*? [Refer to the *Earth Science Reference Tables*.]   (1) 1 m/km (2) 10 m/km   (3) 3 m/km   (4) m/km

Additional Questions

1. Which is a model?   (1) a globe   (2) a ruler   (3) a hand lens (4) a mineral specimen

To answer questions 2 and 3 refer to the diagram below, which shows a satellite in a polar orbit around the earth. Assume that the satellite contains photographic equipment.

2. Which of the following could not be easily determined from data from the satellite?   (1) the thickness of the lithosphere   (2) the circumference of the earth   (3) the shape of the earth (4) the surface area of the earth

3. If the satellite were in a perfectly circular orbit, which of the following statements would be true of its distance from the surface of the earth? (1) it would be constant   (2) it would be greatest at the poles   (3) it would be greatest at the equator   (4) it would sometimes be greatest at the poles and at other times greatest at the equator

4. Which of the following cannot be determined about the earth when the only dimension known is the circumference?   (1) mass (2) volume   (3) diameter   (4) surface area

5. Which makes up most of the earth's surface?   (1) the atmosphere (2) the lithosphere   (3) the hydrosphere

The diagrams below illustrate systems that can be used to determine position on a sphere. Refer to them to answer questions 6 and 7.

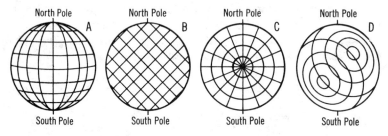

6. Systems of lines like those illustrated above are called   (1) latitude systems   (2) coordinate systems   (3) great circle systems   (4) axis systems

7. Which of the systems illustrated is most like the latitude-longitude system used on the earth?   (1) A   (2) B   (3) C   (4) D

TOPIC IV   The Earth's Motions

APPARENT MOTIONS OF CELESTIAL OBJECTS

CELESTIAL OBJECTS.   A **celestial object** is any object outside the earth's atmosphere. The stars, the sun, the planets, and the moon are examples of celestial objects. The solar system is the sun and the portion of space containing celestial objects that revolve around the sun. See *Solar System Data* in the *Earth Science Reference Tables*.

DAILY MOTION.   Most celestial objects appear to move across the sky from east to west. The paths of these apparent motions are circular, and the motion occurs at a constant rate of approximately 15° per hour. This is called **daily motion.** See Figure 4-1.

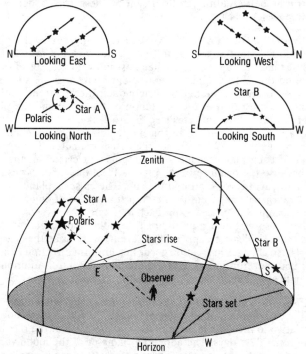

Figure 4-1. The apparent daily motion of the stars. To an observer in the mid-latitudes of the Northern Hemisphere, most stars appear to move from east to west in circular paths, or along parts of circles called **arcs,** at an angular rate of 360° in 24 hours (15° per hour). The center of these circular paths and arcs is very near the star Polaris. The complete circular path can be seen for stars in the northern part of the sky around Polaris. Other stars rise over the eastern horizon and set at the western horizon.

MODELS TO ACCOUNT FOR CELESTIAL OBSERVATIONS. The most obvious way to explain the daily motion of celestial objects is to assume that the earth is stationary and that the celestial objects are revolving around it at a rate of 15° per hour. This is called the *geocentric* ("earth-centered") model. It is the model that was employed by the ancient Greek scientists and that was accepted up to the 16th century. However, the geocentric model fails to explain certain observed **terrestrial motions** (motions related to the earth). In modern times it has been replaced by the *heliocentric* ("sun-centered") model. In this model the earth is rotating on its axis at the rate of 15° per hour and is revolving around the sun once per year. (**Rotation** of a body is the turning of the body on its own axis, in the manner of a spinning toy top. **Revolution,** or revolving, of a body is its movement around another body in a path called an **orbit.**) The heliocentric model accounts for all observed motions of celestial objects and terrestrial objects in a simpler way than the geocentric model. (Additional discussion of these two models will be found on pages 42–44.)

MOTIONS OF THE PLANETS. As seen from the earth, the planets also exhibit daily motion similar to that of the stars, but over extended periods of time they appear to change position with respect to the *star field* around them. This apparent movement relative to the stars is not uniform. According to the heliocentric model, the apparent motion of the planets is the result of the revolution of the earth and the planets around the sun at different rates (see Figure 4-2 on page 31).

The **apparent diameter** of an object is the diameter or size the object appears to have to an observer, not the actual diameter. The apparent diameter of each planet changes in a regular, or cyclic, manner. This occurs because the distance of the planets from the earth varies as the planets and earth revolve around the sun. The closer a planet is to the earth, the larger its apparent diameter, just as a distant object on earth seems to become larger as we approach it.

When observed over a period of time, identifiable features of the planets, such as volcanoes, appear in different locations on the face of the planet. These changes in location occur in a uniform direction in a cyclic manner and therefore indicate that these planets rotate. Additional evidence indicates that all the planets rotate.

MOTIONS OF THE MOON. As viewed from the earth, the moon shares the daily motion of all celestial objects. In addition, it appears to move relative to the stars in an orbit around the earth at the rate of one revolution per 27⅓ days. The apparent diameter of the moon varies in a cyclic manner at the same rate. These observations can be explained by the revolution of the earth and the moon in elliptical orbits around the center of mass of the earth-moon system, while the system as a whole revolves around the sun.* The moon's changes in apparent diameter can be explained by the revolution of the moon around the earth in an elliptical orbit.

---

*Elliptical orbits are discussed on page 37.

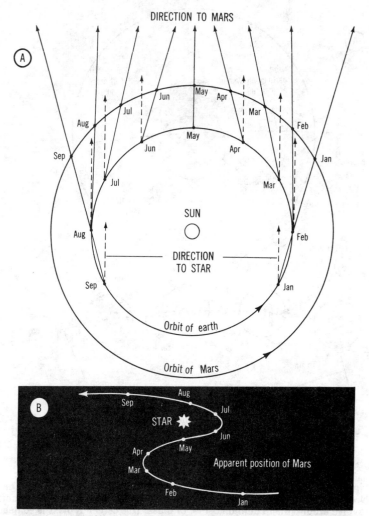

Figure 4-2. Apparent motion of a planet. Diagram A shows successive positions of the earth and Mars along their orbits at monthly intervals during part of a particular year. The dashed arrows show the sight lines from the earth to some fixed star. These sight lines are practically parallel because of the great distance of the star. The solid arrows are the sight lines toward Mars. As the diagram shows, the sight line in January was to the right of the star. From January to March, Mars appeared to move to the left. From March to June it appeared to move to the right. From June to September, it appeared to move to the left again. As a result, the apparent position of Mars changed with respect to the star as shown in Diagram B. Because the orbits of the earth and Mars are tilted with respect to each other, the altitude of Mars with respect to the fixed star also changed. In the following year, a similar motion would be observed, but in a different region of the heavens.

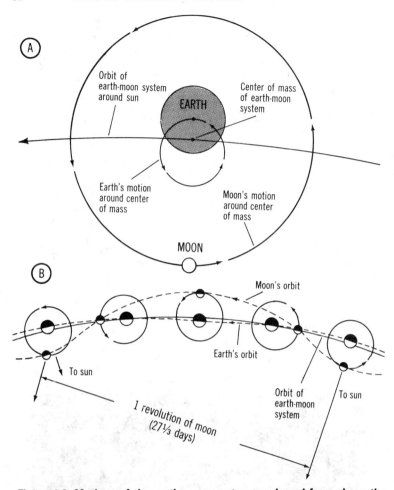

Figure 4-3. Motions of the earth-moon system as viewed from above the North Pole. Diagram A shows that the centers of the earth and the moon revolve around their common center of mass. This center of mass is so near the earth's center that it is actually inside the earth, about 1700 kilometers below the surface. Diagram B shows the motion of the earth-moon system as it revolves around the sun. The oscillations of the orbits are greatly exaggerated. If drawn to scale, the radius of the moon's orbit in Diagram B would be about 1/50 inch.

MOON PHASES. Half of the moon is always receiving light from the sun at any given time. Since the moon revolves around the earth, an observer on the earth sees varying amounts of the lighted half as the moon changes position in its orbit. Those varying amounts of the lighted moon are known as its **phases** (see Figure 4-4).

Because the revolution of the moon around the earth-moon focus is cyclic, the phases of the moon are also cyclic. However, because of the revolution of the earth-moon system around the sun, the cycle of phases is somewhat longer than the time of one revolution of the moon. The period from one full moon to the next is 29½ days, whereas the period of revolution is 27⅓ days. (See Figure 4-5.)

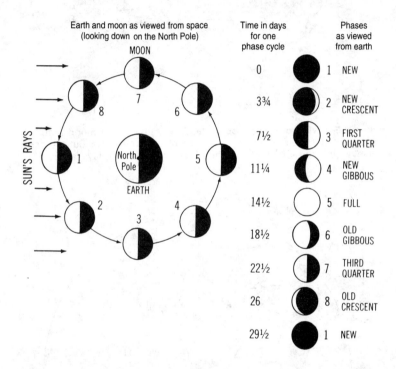

Figure 4-4. The phases of the moon. The half of the moon facing the sun at any given time is illuminated. The other half is dark. An observer on the earth sees varying portions of the illuminated half, depending on the position of the moon in its orbit. The appearance of the phases for eight positions in the cycle is illustrated.

MOTIONS OF THE SUN. Like all other visible celestial objects, the sun also goes through daily motion in the sky. The sun's apparent path, extending from sunrise to sunset, has the shape of an **arc** (part of a circle). As shown in Figure 4-6, that path changes position and length with the seasons. The greater the length of the path, the longer the daylight period. Figure 4-6 also shows how the position of the sun at sunrise and sunset varies predictably with the seasons.

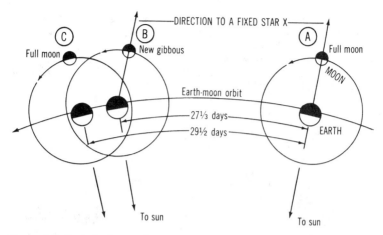

Figure 4-5. Period of one phase cycle. At Position A, the moon is full. At Position B, the moon has made one complete revolution about the earth. However, it is not full until it reaches Position C (about 2 days later).

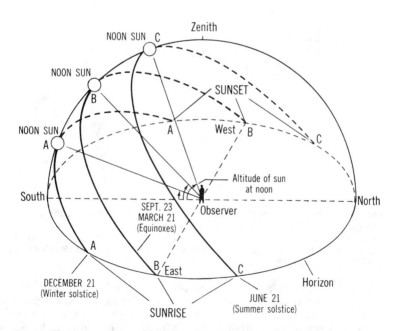

Figure 4-6. The changing path of the sun during the year at 42° north latitude. Note that the path of the sun is longest on June 21, shortest on December 21, and average in length on March 24 and September 23.

ALTITUDE OF THE NOON SUN. The sun always reaches its highest position in the sky at local noon (see page 18). However, the altitude of the sun at noon depends on the time of year and the latitude of the observer. Only between latitudes 23½° N and 23½° S can the noon sun ever be directly overhead. Thus the noon sun is never directly overhead anywhere in the continental United States. Figure 4-7 shows how the latitude at which the noon sun is overhead changes during a year.

If you know the latitude at which the noon sun is directly overhead on a particular date, you can calculate the altitude of the noon sun for any other location on that date. First find the number of degrees between the latitude of the observer and the latitude at which the noon sun is overhead. Then subtract that number of degrees from 90°. For example, to find the altitude of the noon sun at 40° N on December 21, first note that on December 21 the noon sun is directly overhead at 23½° S. The number of degrees between 23½° S and 40° N is 63½° (23½° S to 0°, plus 0° to 40° N). Now subtract 63½° from 90°. The result is 26½°. This is the altitude of the noon sun at 40° N on December 21.

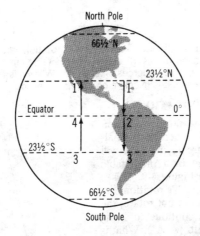

Figure 4-7. The changes in latitude where the sun is directly overhead (at the zenith point) at noon in the course of one year. Point 1—June 21   Point 3—Dec. 21   Point 2—Sept. 23   Point 4—March 21

ROTATION OF THE EARTH

The main motions of the earth are its rotation and its revolution. The earth's rotation is the spinning of the earth on its axis, the imaginary line from the North Pole to the South Pole through the planet. The earth rotates from *west to east* 360 degrees in 24 hours, an angular rate of 15 degrees an hour. The linear rotational speed (as in km/h) of a point on the earth's surface depends on its latitude. As shown in Figure 4-8, the circumference of the circle of rotation increases as you go from the poles toward the equator. Therefore, the linear speed of rotation also increases in the same way.

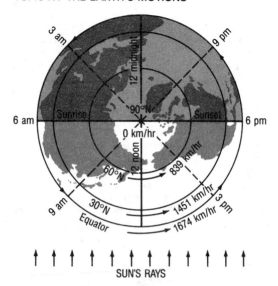

SUN'S RAYS

Figure 4-8. Circles of rotation showing linear rotation rates at different latitudes. Angular rotation rate is 15 degrees per hour at any latitude. Solar time changes 1 hour for each 15 degrees of longitude. When it is noon on the side of the earth facing the sun, it is midnight on the opposite side.

EVIDENCE OF ROTATION

1. The Foucault pendulum. When the Foucault pendulum is allowed to swing freely, its path will appear to change in a predictable way, as shown in Figure 4-9. This is evidence for the earth's rotation, because if the earth did not rotate, the pendulum would continue to swing in the original path (A-A'). What is really happening is that the earth is rotating under the pendulum, so that the pendulum's path appears to change.

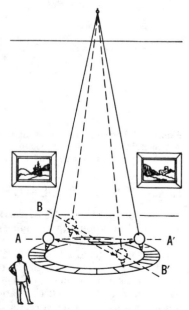

Figure 4-9. Apparent motion of Foucault pendulum. An observer sees a pendulum swing in the direction A-A'. Several hours later, the pendulum has changed its direction of swing to the line B-B'. Actually, the pendulum swings in a fixed direction in space, while the earth, carrying the observer with it, rotates under the pendulum.

2. The Coriolis effect. The **Coriolis effect (Coriolis force)** is the tendency of all particles of matter moving at the earth's surface to be deflected from a straight-line path. The deflection is to the right in the Northern Hemisphere and to the left in the Southern Hemisphere. The deflection occurs because the earth is rotating, and therefore the earth's surface is moving with respect to the path of the particles. The following example can help to explain the effect: You are at the center of a merry-go-round that is rotating counterclockwise. Your friend is near the rim of the merry-go-round. You throw a ball directly at your friend. By the time the ball reaches the rim of the merry-go-round, your friend has been carried to the left. The ball reaches the rim at a point that is now to the right of your friend. With respect to the moving merry-go-round, the ball has been deflected to the right. In a similar manner, ocean currents and winds are deflected with respect to the earth's surface.

REVOLUTION OF THE EARTH

A planet's revolution is its motion around the sun in a path called an **orbit.** The shape of the earth's orbit (and those of the other planets) is a closed curve called an **ellipse.** Within the ellipse are two fixed points called **foci** (singular, **focus**). The sum of the distances between any point on the ellipse and the two foci is a constant; that is, the sum of those distances for one point is equal to the sum for any other point on the curve. The sun is at one of the foci of each planetary orbit in the solar system (see Figure 4-10).

The flattening of an ellipse is measured by its **eccentricity.**

$$\text{eccentricity of an ellipse} = \frac{\text{distance between foci}}{\text{length of major axis}} = \frac{d}{L}$$

If you measure d and L in the ellipse of Figure 4-10 and apply the formula, you should find the eccentricity to be 0.25. As the foci of an ellipse are brought closer together, the ellipse becomes more like a circle and the eccentricity decreases toward zero. The eccentricities of the planets are listed in *Solar System Data* in the *Earth Science Reference Tables.*

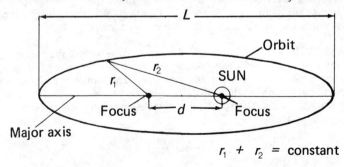

Figure 4-10. An elliptical orbit of a planet with the sun at one focus. Flattening of ellipse and separation of foci are greatly exaggerated.

VARYING DISTANCE OF PLANETS FROM THE SUN.   The elliptical shape of planetary orbits causes the planets to vary in distance from the sun during a revolution. For example, the earth is 147,000,000 kilometers from the sun when closest (**perihelion,** January 3) and 152,000,000 kilometers from the sun when farthest away (**aphelion,** July 4). The difference between these two distances is the same as the distance between the two foci (5,000,000 km). Since this distance is relatively small compared to the major axis (299,000,000 km), the eccentricity of the orbit is small (0.017) and the orbit is very nearly a circle.

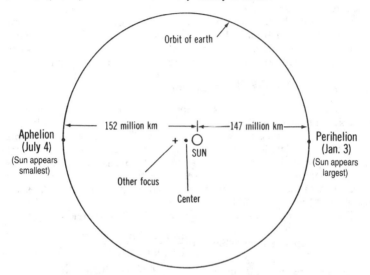

Figure 4-11. The earth's orbit. If drawn exactly to scale, the two foci would be about 1 mm apart.

The changing distance of the earth from the sun causes the sun's apparent diameter to change in a cyclic fashion during the year. The sun's apparent diameter is greatest on January 3 when the sun is closest to the earth and smallest on July 4 when it is farthest.

GRAVITATION.   The earth and the other planets orbit the sun under the influence of **gravitation,** which is the attractive force that exists between any two objects in the universe. The gravitational force is proportional to the product of the masses of the objects and inversely proportional to the square of the distance between their centers. This can be expressed as

$$F \propto \frac{m_1 m_2}{d^2}$$

where m_1 and m_2 are the two masses, d is the distance between the *centers* of the two objects, and *F* is the gravitational force.

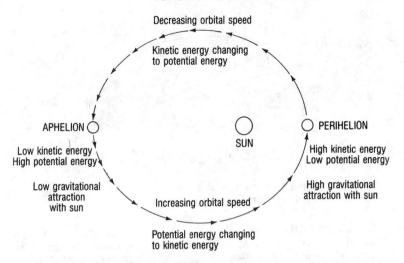

Figure 4-12. Cyclic energy transformation and changing orbital speed of a planet. Elliptical shape of orbit greatly exaggerated.

ORBITAL SPEED. The force of gravitation between the sun and a planet is always pulling the planet toward the sun. The planet does not fall into the sun because of the outward force related to its orbital motion. However, when the planet is in a part of its orbit that is bringing it closer to the sun, the force of gravitation is aiding the motion, causing the planet to speed up. The planet at such times is actually "falling" toward the sun, and some of its **potential energy** (stored energy or energy of position) is converted to **kinetic energy** (energy of motion). As the planet enters the part of its orbit that takes it away from the sun, the force of gravitation tends to oppose this motion. The planet's speed therefore decreases as it "climbs" to a higher distance from the sun. At such times its kinetic energy is being converted back to potential energy. Thus we see that there is a cyclic energy transformation between potential energy and kinetic energy as a planet revolves around the sun (see Figure 4-12). In the case of the earth, it is closest to the sun (perihelion) and has its greatest orbital speed around January 3. It is farthest from the sun (aphelion) and has its least orbital speed around July 4.

Changes in the orbital velocity of a planet always take place according to a definite principle: *An imaginary line connecting the sun and the planet always sweeps through equal areas during equal intervals of time.* This principle is illustrated in Figure 4-13. Since equal areas must be covered in equal intervals of time, the planet's orbital speed must be fastest at perihelion and slowest at aphelion.

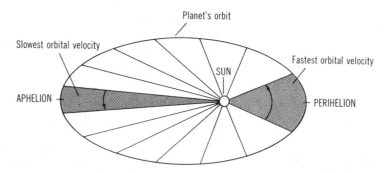

Figure 4-13. A line from the sun to a planet will sweep out each of these equal areas in the same time.

PLANET PERIOD. The **period** of a planet is the amount of time it takes the planet to make one orbit, or revolution, around the sun. This equals one **year** for that planet. The period of a planet is related to the planet's distance from the sun. The closer a planet is to the sun, the smaller its orbit, and the shorter its period. More precisely, for any planet the square of its period of revolution is proportional to the cube of its mean radius of orbit or mean distance from the sun. The relationship is expressed symbolically as $T^2 \propto R^3$, where T is the period of the planet and R is the mean radius of orbit. When T is expressed in earth years and R in Astronomical Units, then $T^2 = R^3$. One Astronomical Unit is the mean distance of the earth from the sun or about 150 million kilometers.

The relationship of period to distance can be easily verified. The period of Mars is 1.88 earth years and its distance from the sun is 1.52 times the average earth distance, or 1.52 Astronomical Units.

$$T = 1.88 \qquad\qquad R = 1.52$$
$$T^2 = 1.88 \times 1.88 \qquad R^3 = 1.52 \times 1.52 \times 1.52$$
$$= 3.53 \qquad\qquad\quad = 3.51$$

Although the results are not equal, they are close enough to verify the relationship.

TIME AND EARTH MOTIONS

THE YEAR. The length of the **year** (or period of revolution) is equal to the time it takes the earth to make one revolution around the sun.

THE APPARENT SOLAR DAY. As explained on page 35, the sun reaches its highest point in the sky once each day at noon. The time it takes for the earth to rotate from noon to noon on two successive days at any fixed location on the earth is called the **apparent solar day**. Time based on the actual motions of the sun in the sky is called **apparent solar time,** or **sundial** time, because a sundial is used to measure it. As shown by Figure 4-14, the apparent solar day is always longer than one rotation of the earth. If the earth did not revolve in its orbit, it would take exactly one rotation to bring a point on the earth into the same position with respect to the sun. However, because the earth does move along its orbit, it has to rotate a small additional amount to bring the point into the same position relative to the sun.

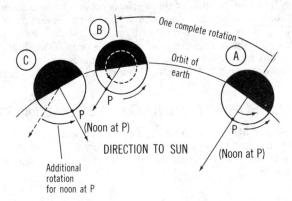

Figure 4-14. Length of apparent solar day. At Position A, it is noon at Point P on the earth. After one complete rotation, the earth is at Position B, but it is not yet noon at Point P. The earth must rotate an additional amount to bring Point P in line with the sun again, when it is again noon at P. The earth will have moved to Position C in this time. (The drawing is exaggerated to make the principle clear.)

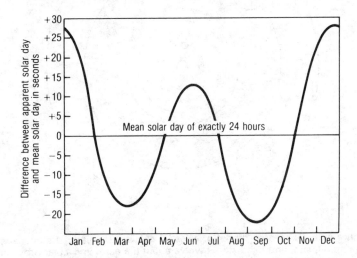

Figure 4-15. Variations in the length of the apparent solar day. This curve shows the difference between the apparent solar day and the mean solar day during the course of one year.

MEAN SOLAR DAY. If the earth moved around the sun in a circular orbit at constant speed, the apparent solar day would have a constant length. However, the earth's speed around the sun varies in a cyclic manner in the course of the year. As a result, the length of the apparent solar day also varies during the year. The inclination of the earth's axis, and the varying curvature of its orbit, also have an effect on the apparent solar day. The combination of these effects produces the cyclic variation in the apparent solar day shown in Figure 4-15. For convenience in timekeeping, a solar day of average length, called the **mean solar day,** has been established. This mean solar day has been divided into exactly 24 hours. On most days of the year, the apparent solar day is either slightly more or slightly less than 24 hours in length (as already noted in Figure 4-15).

MODELS FOR EXPLAINING CELESTIAL MOTIONS

GEOCENTRIC MODEL. An early concept of celestial objects and their motions is the **geocentric model** ("earth-centered"). In this model, the earth is stationary and all celestial objects revolve around it, as shown in Figure 4-16.

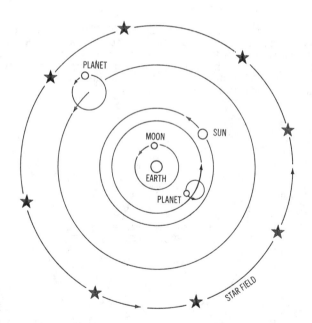

Figure 4-16. Geocentric model of celestial objects. The earth is stationary. The moon, sun, and fixed stars revolve about the earth in circular orbits at different speeds. The planets revolve in small circles, while the centers of these circles move about the earth in circular orbits. These additional circles for the planets are needed to explain their irregular motions.

Some of the major points about the geocentric model are:

1. It explains the apparent motions of celestial objects. The daily motion of the sun and stars is explained by their revolution around the earth. The irregular motion of the planets is explained by assuming that they move in small circular paths in addition to their revolution around the earth.

2. This model does not explain certain terrestrial motions, such as motions of the Foucault pendulum and the Coriolis effect.

3. The need for auxiliary circles for the planetary motions makes the model complex.

HELIOCENTRIC MODEL. The modern concept of the motions of celestial objects is the **heliocentric model** ("sun-centered"). In this model, the earth and the planets revolve around the sun; in addition, the earth rotates on its axis. (See Figure 4-17.)

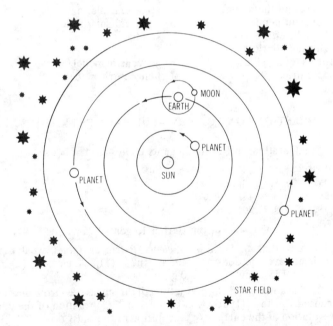

Figure 4-17. Heliocentric model. The sun is stationary. The earth and the planets revolve about the sun at different speeds in elliptical orbits. The stars are stationary and are located at great distances from the solar system.

The major points about the heliocentric model are:

1. It explains the apparent motions of celestial objects. The daily motion is the result of the earth's rotation. The irregular motion of the planets is the result of the revolution of the earth and the planets around the sun at different rates.

2. This model accounts for observed terrestrial motions, such as motions of the Foucault pendulum and the Coriolis effect.

3. It is less complex than the geocentric model because each planet requires only one type of orbital motion.

VOCABULARY

celestial object	perihelion
daily motion	aphelion
arc	gravitation
terrestrial motions	orbital speed (or velocity)
rotation	potential energy
revolution (revolving)	kinetic energy
orbit	period (of revolution)
apparent diameter	year
phases (of the moon)	apparent solar day
Foucault pendulum	apparent solar time
Coriolis effect	sundial
ellipse (elliptical)	mean solar day
focus (of an ellipse)	geocentric model
eccentricity	heliocentric model

QUESTIONS ON TOPIC IV—THE EARTH'S MOTIONS

Questions in Recent Regents Exams (end of book)

June 1995: 6, 12, 14, 17, 57, 58, 61–65
June 1996: 8, 9, 12, 13, 17, 61–65
June 1997: 3, 8–11, 15, 66, 79, 80, 103
June 1998: 7–9, 70, 101, 103

Questions from Earlier Regents Exams

1. According to the *Earth Science Reference Tables,* what is the approximate average density of the earth?   (1) 2.80 g/cm^3 (2) 5.52 g/cm^3 (3) 9.55 g/cm^3 (4) 12.0 g/cm^3

2. The apparent rising and setting of the sun as seen from the earth are caused by the   (1) rotation of the sun   (2) rotation of the earth   (3) revolution of the earth   (4) revolution of the sun

3. How many degrees does the sun appear to move across the sky in two hours?   (1) 15°   (2) 30°   (3) 45°   (4) 60°

4. If the earth rotated from north to south, the North Star would appear to   (1) set in the south   (2) set in the north   (3) move in a circle in the sky   (4) remain stationary

5. If viewed from the earth over a period of several years, the apparent diameter of Mars will   (1) decrease constantly   (2) increase constantly   (3) remain unchanged   (4) vary in a cyclic manner

6. How would a three-hour time exposure photograph of stars in the northern sky appear if the earth did *not* rotate?

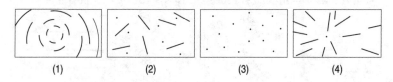

(1)   (2)   (3)   (4)

7. An observer took a time-exposure photograph of Polaris and five nearby stars. How many hours were required to form these star paths?

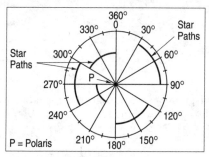

(1) 6   (2) 2   (3) 8   (4) 4

8. A planet was viewed from earth for several hours. The diagrams below represent the appearance of the planet at four different times. The best inference that can be made based on the diagrams is that this planet is (1) tilted on its axis   (2) changing seasons   (3) revolving   (4) rotating

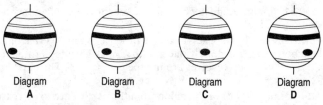

Diagram   Diagram   Diagram   Diagram
   A   B   C   D

Base your answers to questions 9 through 11 on the *Solar System Data Table* in the *Earch Science Reference Tables.*

9. Which planet has traveled around the sun more than once in your lifetime?   (1) Mars   (2) Uranus   (3) Neptune   (4) Pluto

10. On which of the following planets would the Coriolis effect be greatest?   (1) Mars   (2) Mercury   (3) Jupiter   (4) Neptune

11. Which planet has the most eccentric orbit?   (1) Venus   (2) Mars   (3) Saturn   (4) Pluto

Base your answers to questions 12 through 15 on the diagram below. The diagram represents the moon in various positions in its orbit around the earth. Letters *A* through *E* represent five of the moon's positions.

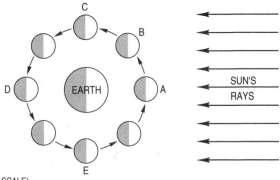

(NOT DRAWN TO SCALE)

12. Which diagram best represents the appearance of the moon to an observer on the earth when the moon is at position *B*?

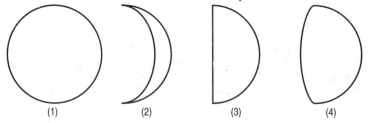

(1)   (2)   (3)   (4)

13. The moon would *not* be visible from the earth when the moon is at position   (1) *A*   (2) *E*   (3) *C*   (4) *D*

14. Why would an observer on earth see a complete cycle of phases of the moon in approximately one month?   (1) The moon rotates on its axis.   (2) The moon revolves around the earth.   (3) The earth rotates on its axis.   (4) The earth revolves around the sun.

15. If the distance of the moon from the earth were to increase, the length of time the moon would take to complete one revolution around the earth would   (1) decrease   (2) increase   (3) remain the same

16. To a person located at 43° north latitude, the sun appears to rise due east on   (1) December 22   (2) March 1   (3) March 21   (4) June 22

17. At noontime on June 21, the sun will make an angle of 23.5° with an observer's zenith if the observer is located at   (1) 0°   (2) 23.5°N   (3) 23.5°S   (4) 66.5°N

18. At 40° north latitude, for how many days a year is the sun in the zenith (directly overhead) position at noon?   (1) 1   (2) 2   (3) 3   (4) 0

19. In New York State, to see the sun at noon, one would look towards the   (1) north   (2) south   (3) east   (4) west

Base your answers to questions 20 through 23 on the diagram below. The diagram represents a plastic hemisphere upon which lines have been drawn to show the apparent paths of the sun on four days at a location in New York State. Two of the days are December 21 and June 21. The protractor is placed over the north-south line.

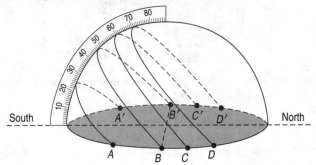

20. On which two dates could the sun have followed path C-C'? (1) October 22 and March 28   (2) September 9 and January 7   (3) January 27 and August 21   (4) May 7 and August 1

21. Which path was recorded on a day that had twelve hours of daylight and twelve hours of darkness?   (1) A-A'   (2) B-B'   (3) C-C'   (4) D-D'

22. Which would be the approximate length of the daylight period for the observer when the sun travels along the entire length of path A-A?'   (1) 9 hours   (2) 12 hours   (3) 15 hours   (4) 18 hours

23. Which observation about the sun's apparent path on June 21 is best supported by the diagram?   (1) The sun appears to move across the sky at a rate of 1° per hour.   (2) The sun's total daytime path is shortest on this date.   (3) Sunrise occurs north of east.   (4) Sunset occurs south of west.

24. According to the diagram below, the time at point *C* is closest to (1) 6 A.M.   (2) 12 noon   (3) 6 P.M.   (4) 12 midnight

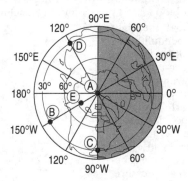

25. A student in New York State obtained the data below by noting the altitude of the sun at noon for three consecutive months.

Month	Altitude of sun at noon
X	27.0°
Y	23.5°
Z	25.5°

The data for month *Y* were obtained during   (1) March   (2) June   (3) September   (4) December

26. As one moves from the poles toward the equator, the linear velocity in km/hr of the earth's surface caused by the earth's rotation   (1) decreases   (2) increases   (3) remains the same

27. The diagram below shows the rotating earth as it would appear from a satellite over the North Pole.

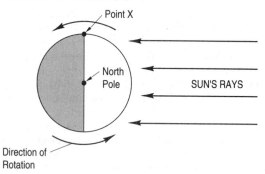

The time at point *X* is closest to   (1) 6 A.M.   (2) 12 noon   (3) 6 P.M.   (4) 12 midnight

28. The Foucault pendulum provides evidence of the earth's   (1) rotation   (2) revolution   (3) precession   (4) inclination

29. The diagram at the right represents a Foucault pendulum in a building in New York State. Points *A* and *A'* are fixed points on the floor. As the pendulum swings for six hours, it will   (1) appear to change position due to the earth's rotation   (2) appear to change position due to the earth's revolution   (3) continue to swing between *A* and *A'* due to inertia   (4) continue to swing between *A* and *A'* due to air pressure

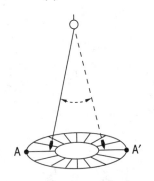

30. Planetary winds do not blow directly north or south because of (1) the Coriolis effect   (2) gravitational force   (3) magnetic force (4) centripetal force

31. The Coriolis effect provides evidence that the earth   (1) has a magnetic field   ((2) has an elliptical orbit   (3) revolves around the sun (4) rotates on its axis

32. Some stars that can be seen in New York State on a summer night cannot be seen on a winter night. This fact is a result of the   (1) rotation of the earth on its axis   (2) rotation of the stars around Polaris   (3) revolution of Polaris around the earth   (4) revolution of the earth around the sun

33. Based on the diagram below and the *Earth Science Reference Tables,* what is the eccentricity of the ellipse shown below?   (1) 1.0 (2) 0.5   (3) 0.25   (4) 0.13

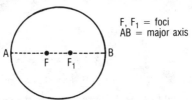

F, F_1 = foci
AB = major axis

34. If the pins in the diagram at the right were placed closer together, the eccentricity of the ellipse being constructed would   (1) decrease (2) increase   (3) remain the same

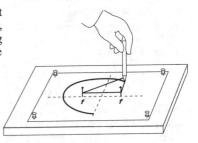

35. Which diagram best approximates the shape of the path of the earth as it travels around the sun?

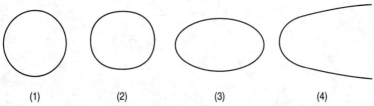

(1)   (2)   (3)   (4)

36. If the distance between the earth and the sun were increased, which change would occur?   (1) The apparent diameter of the sun would decrease.   (2) The amount of insolation received by the earth would increase.   (3) The time for one earth rotation (rotation period) would double.   (4) The time for one earth revolution (orbital period) would decrease.

37. In the Northern Hemisphere, during which season does the earth reach its greatest distance from the sun?   (1) winter   (2) spring   (3) summer   (4) fall

38. If the distance from the earth to the sun were doubled, the gravitational attraction between the sun and earth would become (1) one-fourth as great   (2) one-third as great   (3) twice as great   (4) four times as great

39. In the diagram below, letters A through D represent the locations of four observers on the earth's surface. Each observer has the same mass.

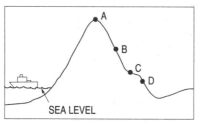

(NOT DRAWN TO SCALE)

The gravitational force is strongest between the center of the earth and the observer at location   (1) A (2) B (3) C (4) D

40. Base your answer to the question on the diagram below, the metric scale in the *Earth Science Reference Tables,* and the equation for gravitational force.

Which would represent the change in gravitational force between sphere 1 and sphere 2 after sphere 2 is moved from position A to position B? (1) ⅙ as great   (2) ½ as great   (3) ⅓ as great   (4) ⅑ as great

41. The diagram below represents a planet in orbit around a star. Which statement best describes how the planet's energy is changing as it moves from point A to point B?

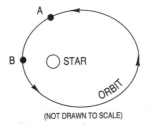

(NOT DRAWN TO SCALE)

(1) Kinetic energy is increasing and potential energy is decreasing.
(2) Kinetic energy is decreasing and potential energy is increasing.
(3) Both kinetic and potential energy are decreasing   (4) Both kinetic and potential energy are increasing.

42. When the orbital velocity of the earth is greatest, what is the season in the Northern Hemisphere?   (1) spring   (2) summer   (3) fall   (4) winter

43. The elliptical shape of the earth's orbit causes   (1) Foucault pendulums to change direction   (2) the earth to have an oblate spheroid shape   (3) the earth's axis to be inclined to its orbit   (4) changes in the orbital velocity of the earth

44. The diagram below shows a planet's orbit around the sun. At which location is the planet's orbital velocity greatest?   (1) *A*   (2) *B*   (3) *C*   (4) *D*

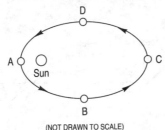

(NOT DRAWN TO SCALE)

45. Planet *A* has a greater mean distance from the sun than planet *B*. On the basis of this fact, which further comparison can be correctly made between the two planets?   (1) Planet *A* is larger.   (2) Planet *A*'s revolution period is longer.   (3) Planet *A*'s speed of rotation is greater.   (4) Planet *A*'s day is longer.

46. The diagram below shows the orbits of planets *A* and *B* in a star-planet system.

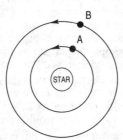

The period of revolution for planet *B* is 40 days. The period of revolution for planet *A* most likely is   (1) less than 40 days   (2) greater than 40 days   (3) 40 days

47. Cities located on the same meridian (longitude) must have the same   (1) altitude   (2) latitude   (3) length of daylight   (4) solar time

48. When does local solar noon always occur for an observer in New York State?   (1) when the clock reads 12 noon   (2) when the sun reaches its maximum altitude   (3) when the sun is directly overhead   (4) when the sun is on the Prime Meridian

49. Upon which frame of reference is time based?   (1) the motions of the earth   (2) the longitude of an observer   (3) the motions of the moon   (4) the real motions of the sun

50. If the earth's rate of rotation decreased, there would be an increase in the   (1) length of the seasons   (2) sun's angle of insolation at noon   (3) number of observable stars at night   (4) length of time for one earth day

51. The graph below shows the time of apparent solar noon (maximum altitude of the sun) and the time of mean solar noon (clock time) during the year. How does the time of apparent solar noon compare with the time of mean solar noon?

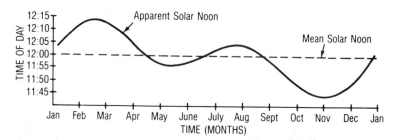

(1) Apparent solar noon always occurs after mean solar noon.   (2) Apparent solar noon and mean solar noon never occur at the same time. (3) The time of apparent solar noon changes with the seasons.   (4) The time of mean solar noon changes during the year.

52. For what reason did the heliocentric model of the universe replace the geocentric model of the universe?   (1) The geocentric model no longer predicted the positions of the constellations.   (2) The geocentric model did not predict the phases of the moon.   (3) The heliocentric model provided a simpler explanation of the motions of the planets.   (4) The heliocentric model proved that the earth rotates.

53. In the geocentric model (the earth at the center of the universe), which motion would occur?   (1) The earth would revolve around the sun.   (2) The earth would rotate on its axis.   (3) The moon would revolve around the sun.   (4) The sun would revolve around the earth.

54. Which observation can *not* be explained by a geocentric model? (1) Stars follow circular paths around Polaris.   (2) The sun's path through the sky is an arc.   (3) A planet's apparent diameter varies. (4) A freely swinging pendulum appears to change direction.

ENERGY. All earth processes result from the transfer of energy. **Energy** is the ability to do work.

ELECTROMAGNETIC ENERGY

Electromagnetic energy is energy that is radiated in the form of waves. All matter gives off electromagnetic energy unless it is at **absolute zero,** which is theoretically the lowest possible temperature and one at which the particles of matter have no motion. Electromagnetic energy has **transverse wave** properties, which means that the waves vibrate at right angles to the direction in which they are moving (see Figure 5-1). As an example, when a rope tied to a solid object is shaken, the particles making up the rope move up and down, but a transverse wave travels along the length of the rope.

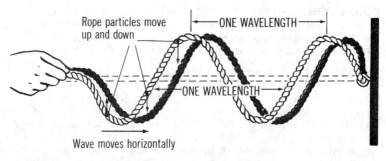

Figure 5-1. A transverse wave in a rope at two moments a short time apart. As the particles of the rope move up and down, the form of the wave moves to the right. In an electromagnetic wave, there are no moving particles. Instead, there are varying electric and magnetic forces at right angles to the direction of propagation of the wave.

The different types of electromagnetic energy are distinguished by their different *wavelengths.* The **wavelength** is the distance from one crest of the wave to the next crest or between corresponding points on successive cycles. This group of electromagnetic radiations is known as the **electromagnetic spectrum.** See the Electromagnetic Spectrum diagram in the *Earth Science Reference Tables.*

INTERACTION OF ELECTROMAGNETIC ENERGY WITH AN ENVIRONMENT. When electromagnetic energy interacts with a material, it can be   (1) **refracted,** or bent in its passage through the material; (2) **reflected,** or bounced off the material;   (3) **scattered,** or refracted and/or reflected in various directions; or   (4) **absorbed,** or taken into the material.

SURFACE PROPERTIES AND ABSORPTION. The characteristics of a surface determine the quantity and type of electromagnetic energy that can be absorbed. For example, the darker and rougher the surface of a material, the more visible light will be absorbed. The better a material is at absorbing electromagnetic energy, the better it also is at giving off or radiating electromagnetic energy. Thus a dark-colored object will heat up fast in sunlight but will also cool off fast after sunset. A material that does not absorb electromagnetic energy well may reflect it well. Light-colored smooth surfaces do not absorb visible light well, but they are good reflectors of visible light.

METHODS OF ENERGY TRANSFER

The three methods of energy transfer are *conduction, convection,* and *radiation.* Conduction and convection refer to the transfer of heat energy in matter (solids, liquids, and gases). Radiation refers to the transfer of energy by electromagnetic waves in space and in matter.

CONDUCTION. **Conduction** is the transfer of heat energy from atom to atom or molecule to molecule through contact when atoms or molecules collide. Conduction is most effective in solids, because the atoms or molecules are closer together than in gases and liquids.

CONVECTION. **Convection** is the transfer of heat by movements in fluids (gases and liquids) caused by *differences in density* within the fluid. Warmer portions of the fluid usually have lower density and tend to rise above cooler portions. The result is a circulatory motion called a **convection cell** or **convection current** which transfers heat energy from one place to another. (See Figure 5-2.) Convection currents transfer heat through the earth's atmosphere, hydrosphere, and perhaps in parts of the mantle zone below the lithosphere.

RADIATION. In **radiation,** no medium is needed to transfer the transverse waves that carry electromagnetic energy; such energy can radiate from its source across empty space. This is the method by which the sun's electromagnetic energy gets through space to the other objects in the solar system. The higher an object's temperature, the more electromagnetic energy it gives off. (See the diagram *Electromagnetic Spectrum* in the *Earth Science Reference Tables.*)

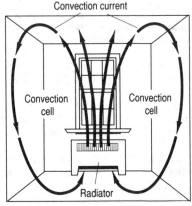

Figure 5-2. Heating a room by convection.

TRANSFER OF ENERGY WITHIN A CLOSED SYSTEM

SOURCE AND SINK. Energy moves away from an area of high concentration, called a **source,** to an area of low concentration, called a **sink.** The energy will continue to move from the source to the sink until their energies are equal, establishing a **dynamic equilibrium.** At dynamic equilibrium an object loses and gains equal amounts of energy. If the dynamic equilibrium is between all forms of energy, then the temperature of the object or system will remain constant.

CONSERVATION OF ENERGY. In any energy system that is closed, or isolated, the total amount of energy remains the same. Thus the amount of energy lost by a source must equal the amount of energy gained by a sink. This concept, that energy is not created or destroyed but stays the same in amount in a closed system, is called the principle of **conservation of energy*.**

KINETIC AND POTENTIAL ENERGY. An object in motion has a kind of energy called **kinetic energy.** Objects or systems can also have a kind of energy called **potential energy,** which is energy related to position or state. It may be thought of as "stored" energy. Either of these two kinds of energy can be transformed into the other. For example, water at the top of a waterfall has potential energy because of its position with respect to the earth's center of mass. As the water falls to a lower level, some of its potential energy is transformed to kinetic energy, resulting in an increase in speed. In a swinging pendulum, there is a cyclic exchange of kinetic and potential energy. At the top of its swing, the pendulum has zero kinetic energy and maximum potential energy. As it passes through the low point of its swing, it has its maximum kinetic energy and minimum potential energy.

FRICTION AT INTERFACES. Transformations of energy occur when there is *friction.* For example, as a stream flows, some of its kinetic energy is transformed into heat energy at the interface between the stream and its bed. This transformation occurs because of friction between the moving water and the bed.

WAVELENGTH CHANGE. A good absorber of electromagnetic energy is also a good radiator. Often when electromagnetic energy is absorbed by an object, it is reradiated at a longer wavelength. This is very common at the earth's surface, where the relatively short-wavelength ultraviolet and visible radiations from the sun are absorbed and reradiated as longer-wavelength infrared radiations.

---

*Einstein's theory of relativity states that mass (or matter) and energy are equivalent, and that one can be converted to the other. The release of energy by radioactivity is an example of the conversion of mass to energy. The energy radiated by the sun and the other stars is also generated by the conversion of mass to energy within the stars.

TEMPERATURE AND HEAT

TEMPERATURE.   According to the modern theory of matter, the particles of every material are in continuous, random motion. The particles therefore have kinetic energy. At any moment, some of the particles have large kinetic energies, some have small kinetic energies, and some have intermediate kinetic energies. **Temperature** is a measure of the *average kinetic energy* of the particles of a body of matter. The greater the average kinetic energy of the particles, the higher the temperature of the body. If two bodies have the same temperature, the average kinetic energy of their particles is also the same. This is true regardless of the substances in the bodies.

TEMPERATURE SCALES.   The human senses respond to temperature by the sensations called "hot" and "cold." Temperature can be measured more accurately by instruments called *thermometers.* Thermometers indicate temperature on a scale marked off in *degrees.* There are several different temperature scales in use, with different sizes of degrees and different zero points. Relationships among three common temperature scales are given in the *Earth Science Reference Tables.*

HEAT.   If two bodies at different temperatures are placed in contact, some of the kinetic energy of the particles of the hotter body will be transferred to the particles of the cooler body. As a result, the particles of the cooler body will acquire more kinetic energy on the average, and its temperature will therefore rise. The particles of the hotter body will lose kinetic energy and its temperature will fall. This transfer of energy will continue until both bodies have the same temperature.

Energy that is transferred from a hotter body to a cooler one because of the difference in temperature is called **heat energy** or **thermal energy,** or more simply **heat.** For example, if a stone at a temperature of 80°C is placed in a container of water at 20°C, heat will flow from the stone into the water until both arrive at the same temperature. The heat transfer in this example occurs by conduction and convection.

QUANTITY OF HEAT.   Since an increase in temperature requires an increase in kinetic energy of the particles of a body, it takes twice as much heat energy to raise the temperature of a body 2° as to raise it 1°. It also takes twice as much heat energy to raise the temperature of 2 grams of a substance 1° as to raise the temperature of 1 gram of the same substance by this amount. In general, the quantity of heat involved in a temperature change of a given substance is directly proportional to the amount of temperature change and the mass of the substance.

When a substance "cools off," that is, when its temperature decreases, it releases the same amount of heat energy that was needed to raise its temperature by the same number of degrees. This is an example of the conservation of energy as applied to heat.

THE CALORIE.   The relationship between quantity of heat, mass of material, and temperature change can be used to define a unit of heat called the *calorie*. For this purpose, water is used as the standard material. One **calorie** is defined as the quantity of heat needed to raise the temperature of one gram of liquid water by one degree Celsius. Since one calorie is needed for each gram of water and for each degree Celsius of temperature change, the number of calories needed to change the temperature of a given mass of water by a given number of degrees can be calculated by the following formula:

No. of calories =
mass of water in grams × temperature change in °C × 1 cal/g/°C

Example

How many calories of heat must be added to 10 grams of liquid water to raise its temperature 15°C?

Solution:

$$\text{No. of cal} = 10 \text{ g} \times 15°C \times 1 \text{ cal/g/°C}$$
$$= 150 \text{ cal}$$

SPECIFIC HEAT. It takes one calorie to raise the temperature of one gram of water one degree Celsius. It takes only about 0.2 calorie (⅕ as much) to raise the temperature of one gram of a typical rock one degree Celsius. The quantity of heat needed to raise the temperature of one gram of any substance one degree Celsius is called the **specific heat** of that substance. The specific heat of the rock is 0.2 cal/g/°C. The specific heat of water is 1.0 cal/g/°C. Liquid water has the highest specific heat of naturally occurring substances. All other naturally occurring substances have a specific heat less than 1 (see the *Earth Science Reference Tables*). Therefore, for equal masses, the same amount of heat causes them to heat up or cool off faster than water.

HEAT LOST OR GAINED. The amount of heat lost or gained by a substance (in calories) equals the mass (in grams) times the temperature change (in °C) times the specific heat of the substance. (See *Equations and Proportions* in the *Earth Science Reference Tables*.)

Example

How much heat is needed to heat 15 grams of granite from 15°C to 20°C if the specific heat of the granite is 0.19 cal/g/°C?

Solution:

$$\text{Temperature change} = 20°C - 15°C = 5°C$$
$$\text{No. of cal} = 15 \text{ g} \times 5°C \times 0.19 \text{ cal/g/°C}$$
$$= 14.25 \text{ cal}$$

HEAT ENERGY AND CHANGES OF PHASE

LATENT HEAT. Matter may exist in the solid, liquid, or gaseous phase. As long as a material is in a single one of these phases, its temperature rises as heat is added to it. If, however, the material begins to *change* phase, for example, from solid to liquid, its temperature *remains the same* as heat is added to it. During the phase change, the added heat energy is not increasing the kinetic energy of the molecules, and therefore the temperature does not change. The added heat energy is being converted to a kind of *potential energy* called **latent heat.**

HEAT LOST OR GAINED IN PHASE CHANGE.   When the change of phase is from a solid to a liquid or from a liquid to a gas, latent heat must be gained by the substance. When the phase change is from a gas to a liquid or from a liquid to a solid, latent heat must be lost by the substance. The heat gained (or lost) in a phase change is equal to the product of the mass times the latent heat (change in potential energy) per unit mass. The latent heat varies with the particular substance and type of phase change.

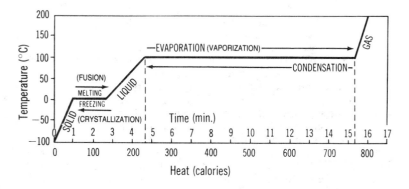

Figure 5-3. Heating curve for water. The graph shows the temperature change of one gram of water as heat is added at a constant rate (50 calories per minute). If read from right to left, the graph is the corresponding cooling curve.

LATENT HEAT AND PHASE CHANGE OF WATER.   Figure 5-3 shows how the temperature of a fixed amount of water changes as it is heated at a constant rate from ice at − 100°C to gaseous water (water vapor) at 200°C. The temperature is plotted against time in minutes (upper scale of the graph) and the corresponding amount of heat added (lower scale of the graph). It can be seen that the temperature remains constant at 0°C for almost 2 minutes while the water is changing from the solid phase (ice) to the liquid phase. There is another constant-temperature interval of more than 10 minutes at 100°C while the water is changing from the liquid phase to the gaseous phase. The reason for these intervals of constant temperature is that the heat being added at those times is being changed to latent heat (potential energy). The latent heat for the change from solid to liquid water (heat of fusion) is 80 calories per gram; for the change from liquid to gas, the latent heat (heat of vaporization) is 540 calories per gram.

THE EARTH'S ENERGY SUPPLY

SOLAR ENERGY. Most of the energy of the earth comes from electromagnetic radiations from the sun. The sun gives off (radiates), and the earth receives, a wide range of electromagnetic energy radiations of various wavelengths. This solar electromagnetic spectrum includes X rays, ultraviolet rays, visible light, and infrared rays. Of all the types of electromagnetic radiations from the sun, the one of greatest intensity is visible light (see Figure 6-1 on page 65).

RADIOACTIVE DECAY. The atoms of certain elements, such as uranium and radium, are unstable and tend to break down into more stable atoms of other elements. This natural, spontaneous breakdown of unstable atoms is called **radioactivity** or **radioactive decay.** It is a process that releases energy and that continues at a constant rate for any particular radioactive element, unaffected by changes in temperature, pressure, or other environmental conditions. Radioactive decay of elements within the earth is an additional source of energy for many earth processes, such as mountain building and volcanic activity.

VOCABULARY

energy	source
electromagnetic energy	sink
absolute zero	dynamic equilibrium
transverse wave	conservation of energy
wavelength	kinetic and potential energy
electromagnetic spectrum	temperature
refraction	heat energy
reflection	calorie
scattering	specific heat
absorption	phase change
conduction	latent heat
convection cell or	radioactivity or radioactive
convection current	decay
radiation	

QUESTIONS ON TOPIC V—ENERGY IN EARTH PROCESSES

Questions in Recent Regents Exams (end of book)

June 1995: 4, 13, 15, 16, 68
June 1996: 11, 15, 16, 19, 68, 69, 105
June 1997: 12–14, 16, 91–95
June 1998: 2, 13, 14, 67, 68, 76, 102

Questions from Earlier Regents Exams

1. At which temperature would an object *not* radiate electromagnetic energy?   (1) 100°C   (2) 0° Fahrenheit   (3) 0° Celsius   (4) absolute zero

2. The various forms of electromagnetic energy are distinguished from one another by their   (1) temperature   (2) wavelengths   (3) longitudinal wave properties   (4) speed of travel

3. As electromagnetic energy from a heat source interacts with its surroundings, it is being absorbed and   (1) reflected, only   (2) refracted, only   (3) scattered, only   (4) reflected, refracted, and scattered

4. Which would absorb the most solar radiation, if you assume that each covers an equal geographic area?   (1) a freshwater lake   (2) a snow field   (3) a sandy beach   (4) a forest

5. Which type of surface would most likely be the best reflector of electromagnetic energy?   (1) dark-colored and rough   (2) dark-colored and smooth   (3) light-colored and rough   (4) light-colored and smooth

6. An object that is a good radiator of electromagnetic waves is also a good   (1) insulator from heat   (2) reflector of heat   (3) absorber of electromagnetic energy   (4) refractor of electromagnetic energy

7. Which process transfers heat from one part of a solid object to another part of the object by direct contact when molecules collide? (1) radiation   (2) advection   (3) conduction   (4) convection

8. By which method is energy transferred by density differences? (1) absorption   (2) conduction   (3) convection   (4) radiation

9. Water is being heated in a beaker as shown below.

Heat Source

Which drawing shows the most probable movement of water in the beaker due to the heating?

(1)   (2)   (3)   (4)

10. At which temperature will an object radiate the greatest amount of electromagnetic energy? [Refer to the *Earth Science Reference Tables.*] (1) 0° Fahrenheit   (2) 5° Celsius   (3) 10° Fahrenheit   (4) 230 Kelvins

11. What method of energy transfer requires no medium for transfer? (1) conduction   (2) convection   (3) advection   (4) radiation

12. An ice cube is placed in a glass of water at room temperature. Which heat exchange occurs between the ice and the water within the first minute?   (1) The ice cube gains heat and the water loses heat.   (2) The ice cube loses heat and the water gains heat.   (3) Both the ice cube and the water gain heat.   (4) Both the ice cube and the water lose heat.

13. An example of a heat sink is   (1) a glacier on a summer day   (2) magma from an erupting volcano   (3) steam from heated ground water (4) an ocean current beginning at the equator

14. The environment is in dynamic equilibrium when it is gaining (1) less energy than it is losing   (2) more energy than it is losing   (3) the same amount of energy as it is losing

15. In a closed system, the amount of energy lost by a heat source (1) is less than the amount of energy gained by heat sink   (2) is greater than the amount of energy gained by a heat sink   (3) equals the amount of energy gained by a heat sink

Questions 16–20 refer to the diagram below. A student is using the apparatus shown to perform an investigation. The two calorimeters contain equal amounts of water, and the metal bar is touching the water inside each calorimeter. At the beginning of the investigation, the temperature of the water was 100°C in calorimeter *A* and 20°C in calorimeter *B*. The room temperature is 20°C.

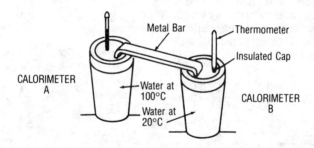

16. At the start of the investigation, which could be considered a heat source?   (1) the water in calorimeter *A*   (2) the water in calorimeter *B*   (3) the air surrounding the calorimeters   (4) the metal bar between the calorimeters

17. If this were a closed system, what would be the temperature when the system reaches equilibrium?   (1) 100°C   (2) 75°C   (3) 60°C   (4) 40°C

18. Which conclusion should the student make after actually performing this investigation?   (1) The energy gained by the cold water equaled the energy lost by the hot water.   (2) The energy gained by the cold water was less than the energy lost by the hot water.   (3) The change in temperature of the cold-water thermometer equaled the change in temperature of the hot-water thermometer.   (4) Energy was transferred between the two calorimeters primarily by radiation.

19. Which procedure would increase the amount of heat energy that is actually gained by calorimeter *B*?   (1) increasing the length of the metal bar   (2) increasing the thickness of the metal bar   (3) circulating air over the metal bar   (4) placing insulation around the metal bar

20. Which graph best represents the probable relationship between the temperatures of the two calorimeters and the time for this heat transfer investigation?

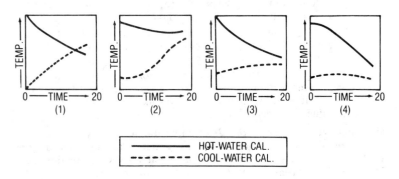

```
HOT-WATER CAL.
COOL-WATER CAL.
```

21. During a volcanic eruption, a rock is thrown upward into the air from point *A* to point *B* as shown in the diagram at the right. Which graph below best represents the relationship between the height of the rock and its potential energy (*PE*) as it rises?

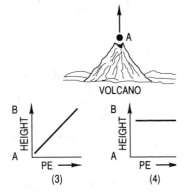

VOLCANO

22. How do the wavelengths of electromagnetic energy absorbed by earth materials compare to the wavelengths reradiated by earth materials?   (1) The reradiated wavelengths are shorter.   (2) The reradiated wavelengths are longer.   (3) The absorbed and reradiated wavelengths are the same.

23. The temperature of an object is determined by the   (1) average kinetic energy of its molecules   (2) average potential energy of its molecules   (3) total kinetic energy of the object   (4) total potential energy of the object

24. According to the *Earth Science Reference Tables*, at which temperature will ice melt under normal conditions?   (1) 0 K   (2) 32 K   (3) 212 K   (4) 273 K

25. According to the *Earth Science Reference Tables*, how many calories of heat energy must be added to 5 grams of liquid water to change its temperature from 10°C to 30°C?   (1) 5 cal   (2) 20 cal   (3) 100 cal   (4) 150 cal

26. Equal masses of lead, granite, basalt, and water at 5°C are exposed to equal quantities of heat energy. Which would be the first to show a temperature rise of 10°C?   (1) lead   (2) granite   (3) basalt   (4) water

27. According to the *Earth Science Reference Tables*, how many calories of heat energy must be added to 10 grams of iron to raise its temperature 10°C?   (1) 0.11 cal   (2) 1.1 cal   (3) 11 cal   (4) 110 cal

28. Based on the *Earth Science Reference Tables*, what is the total amount of energy gained by a 100-gram piece of basalt when it is heated from 20.0°C to 60.0°C?   (1) 40.0 calories   (2) 800. calories   (3) 4,000 calories   (4) 6,000 calories

29. The change from the vapor phase to the liquid phase is called (1) evaporation   (2) condensation   (3) precipitation   (4) transpiration

30. Water vapor crystallizes in the atmosphere to form snowflakes. Which statement best describes the exchange of heat energy during this process?   (1) Heat energy is transferred from the atmosphere to the water vapor.   (2) Heat energy is released from the water vapor into the atmosphere.   (3) Heat energy is transferred equally to and from the water vapor.   (4) No heat energy is exchanged between the atmosphere and the water vapor.

31. Two identical towels are hanging on a clothesline in the sun. One towel is wet; the other is dry. The wet towel feels much cooler than the dry towel because the   (1) dry towel receives more heat energy from the sun   (2) dry towel has more room for heat storage than the wet towel (3) presence of water in the wet towel requires an additional amount of heat energy to bring about a temperature change   (4) water in the wet towel prevents absorption of heat energy

32. Which process results in a release of latent heat energy? (1) melting of ice   (2) heating of liquid water   (3) condensation of water vapor   (4) evaporation of water

33. Water loses energy when it changes phase from   (1) liquid to solid   (2) solid to liquid   (3) liquid to gas   (4) solid to gas

Base your answers to question 34 through 40 on the *Earth Science Reference Tables* and the graph below. The graph shows the temperatures recorded when a sample of water was heated from − 100°C to 200°C. The water received the same amount of heat every minute.

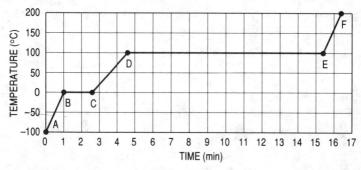

34. For the time on the graph represented by the line from point *B* to point *C*, the water was   (1) freezing   (2) melting   (3) condensing (4) boiling

35. At which point in time would most of the water be in the liquid phase?   (1) 1 minute   (2) 14 minutes   (3) 16 minutes   (4) 4 minutes

36. How many calories were required to change 10. grams of liquid water at point *D* to water vapor at point *E*?   (1) 500 cal   (2) 800 cal   (3) 1,000 cal   (4) 5,400 cal

37. What is the rate of temperature change between points *C* and *D*?   (1) 10 C°/min   (2) 25 C°/min   (3) 50 C°/min   (4) 150 C°/min

38. The greatest amount of energy was absorbed by the water between points   (1) *A* and *B*   (2) *B* and *C*   (3) *C* and *D*   (4) *D* and *E*

39. During which time interval was the rate of temperature change the greatest?   (1) *A* to *B*   (2) *B* to *C*   (3) *C* to *D*   (4) *D* to *E*

40. What is the most probable explanation for the constant temperature between points *D* and *E* on the graph?   (1) The added heat was radiated as fast as it was absorbed.   (2) The added heat was lost to the surroundings.   (3) The added heat changed liquid water to water vapor.   (4) The added heat changed water vapor to liquid water.

41. What is the major source of energy for the earth?   (1) electrical storms   (2) radioactive decay of earth materials   (3) the sun   (4) thermal currents in the mantle

42. The source of energy for the high temperatures found deep within the earth is   (1) tidal friction   (2) incoming solar radiation   (3) decay of radioactive materials   (4) meteorite bombardment of the earth

Insolation and the Seasons

INSOLATION

SOLAR RADIATION. As stated on page 53, every body of matter that is not at absolute zero radiates electromagnetic energy. This energy is radiated over a portion of the electromagnetic spectrum, with more of the energy radiated at some wavelengths than at others. The rate at which energy is radiated is called the *intensity* of the radiation. The higher the temperature of the matter, the shorter the wavelength at which the maximum intensity of radiation occurs. At the temperature of the sun, the maximum intensity of radiation occurs in the range of visible wavelengths.

INSOLATION. Insolation (INcoming SOLar radiATION) is the portion of the sun's radiation that is received by the earth. Figure 6-1 shows how the intensity of insolation varies with wavelength. The maximum intensity occurs in the visible range, but about 50% of the total energy received is infrared, a type of long-wave radiation.

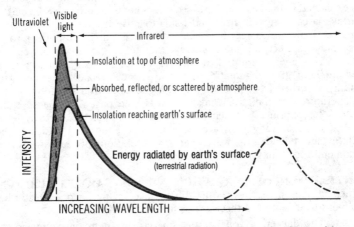

Figure 6-1. Intensity of insolation received and radiated by the earth's surface. Total amount of energy is proportional to the area under the curves. About 50% of the insolation received at the surface is in the infrared range of wavelengths, but 100% of the radiated energy is in this range. (The terrestrial radiation curve is not drawn to the same scale as the insolation curves. If drawn to the same scale, it would be much flatter and would extend much farther to the right.)

EFFECTS OF THE ATMOSPHERE ON INSOLATION. Figure 6-1 also shows that the insolation reaching the earth's surface is different from the insolation entering the earth's upper atmosphere because of *absorption, reflection,* and *scattering* by the atmosphere. Some of the specific effects of the atmosphere on insolation are as follows:

1. Nearly all the ultraviolet radiation is absorbed by the atmosphere.

2. Most of the visible light passes through the atmosphere because the atmosphere is transparent to it.

3. Much infrared radiation is absorbed by water vapor (gaseous water) and carbon dioxide.

4. When clouds are present, much of the incident (incoming) insolation is reflected. For the earth as a whole on a typical day, it is estimated that about 25% of incident insolation is reflected by the cloud cover.

5. *Aerosols* cause random reflection, or scattering, of insolation. **Aerosols** are finely dispersed solids and liquids suspended in air. The dispersed materials include dust, ice crystals, liquid water droplets, and various air pollutants. As the concentration of aerosols increases, the scattering of insolation also increases, thus reducing the amount of insolation that reaches the earth's surface.

ABSORPTION AND RADIATION OF INSOLATION.   When insolation reaches the earth's surface, some of it is reflected and some is absorbed, as explained in the next section. Insolation that is absorbed by the earth's surface is converted to heat and tends to raise the temperature of the surface. However, since the earth's surface is not at absolute zero, it also radiates electromagnetic energy, thus tending to lower its temperature. This energy emitted at the earth's surface is called **terrestrial radiation,** or ground radiation. Over a period of time, the amount of energy absorbed from insolation is equal to terrestrial radiation, and a temperature (or radiative) balance exists (see page 72). Because the earth's temperature is much lower than the sun's, the earth radiates energy at much longer infrared wavelengths. Figure 6-1 indicates the relationship between the wavelengths of insolation and terrestrial radiation emitted from the earth's surface.

INTERACTION OF INSOLATION AT THE EARTH'S SURFACE. When insolation reaches the earth's surface, many factors affect what happens to it. Some of these are as follows:

1. The darker or rougher the surface, the more insolation is absorbed and the less is reflected.

2. Ice and snow reflect almost all of the insolation, keeping areas with ice and snow cooler than they would otherwise be.

3. The altitude of the sun in the sky determines the **angle of insolation,** that is, the angle at which insolation will hit the earth's surface. The lower the angle of insolation, the more insolation is reflected and the less is absorbed.

4. The evaporation of water and the melting of ice and snow transform insolation into latent potential energy which does not raise the temperature of the surface. The more evaporation or melting occurring in an area, the smaller will be the amount of insolation available to heat the area; thus the temperatures will be lower than would otherwise be the case.

LAND AND WATER HEATING.   With equal amounts of insolation, an area consisting largely of water will heat up and cool off more slowly than an equal area of land. There are several reasons for this:

   1. Water has a much higher specific heat than land.

   2. Water is highly transparent to insolation: therefore the insolation is absorbed to a greater depth.

   3. Convection can occur in water, thus distributing the absorbed energy through a much larger volume than is the case with land.

THE GREENHOUSE EFFECT.   Although the atmosphere allows much of the insolation entering it to pass through to the earth's surface, it absorbs most of the energy radiated by the surface. The reason for this is that the longer infrared radiations are absorbed to a large extent by the carbon dioxide and water vapor in the atmosphere. This absorption warms the atmosphere and causes it to act as a heat "blanket," which reduces the loss of energy to outer space and makes the earth's surface warmer than it would otherwise be. This process by which the atmosphere transmits short-wave energy and absorbs radiated long-wave energy (preventing its immediate escape into space) is called the **greenhouse effect.**

   In recent years there has been concern that the greenhouse effect may be causing an increase in earth temperatures. It is thought that release of carbon dioxide into the atmosphere (from the burning of fuels) is increasing the greenhouse effect and heating the planet. This heating could result in environmental problems such as rising sea levels caused by melting glaciers.

VARIATIONS OF INSOLATION

INTENSITY OF INSOLATION.   The rate at which solar energy is received by a given area per unit of time is the **intensity of insolation.** It can be measured in calories per square meter per second.

EFFECT OF ANGLE OF INSOLATION ON INTENSITY AND TEMPERATURE.   When insolation is perpendicular to a surface (that is, striking the surface at an angle of 90°) its intensity is a maximum, because the insolation is concentrated in the smallest possible area. As the angle of insolation decreases from 90° toward zero, the same amount of insolation is spread out over greater and greater areas, as shown by Figure 6-2 (page 68). Therefore the intensity of insolation becomes smaller as the angle becomes smaller. At a high angle of insolation (and therefore higher intensity of insolation), the heating effect of insolation is greater and temperatures are higher.

EARTH'S SHAPE AND INTENSITY OF INSOLATION. The sun's energy reaches the earth as a bundle of parallel rays. If the earth were flat and perpendicular to these rays, the intensity of insolation would be the same everywhere on the earth's surface. However, the earth's surface is nearly a sphere. As a result, at any given time there is just one place where the insolation is perpendicular to the surface; at all other places, the angle of insolation is less than 90°. Therefore the intensity of insolation varies over the earth because of the curvature of its surface.

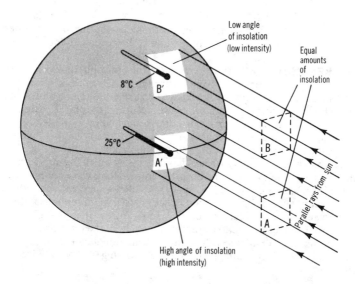

Figure 6-2. Because of the spherical shape of the earth, the parallel rays from the sun strike different parts of the surface at different angles. In high latitudes, where the angle is lower, the same amount of insolation spreads out over a larger area, thus resulting in a lower intensity than where the angle is higher.

INTENSITY OF INSOLATION IN RELATION TO LATITUDE. As shown in Fig. 6-3, insolation is perpendicular to the earth's surface at the equator at each of the **equinoxes** (March 21 and September 23). The rays of **perpendicular insolation** are known as **direct** or **vertical rays.** At each equinox, the intensity of insolation is maximum at the equator, where the direct rays strike, and it decreases with increasing latitude. At the **summer solstice** (June 21), the direct rays strike the earth at 23½°N latitude (the Tropic of Cancer). This is then the latitude of maximum intensity, and the intensity is less at any other latitude. At the **winter solstice** (December 21), the direct rays and the maximum intensity are at 23½°S latitude (the Tropic of Capricorn).

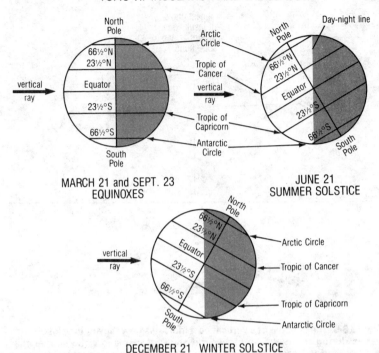

Figure 6-3. Intensity of insolation in relation to latitude. At any given time, intensity of insolation is a maximum at the latitude of the vertical rays, and is less at other latitudes.

INTENSITY OF INSOLATION IN RELATION TO SEASON. As the earth travels along its orbit around the sun, the angle of insolation at any given latitude varies with the seasons, depending on how far from the direct rays the given latitude is. The seasonal change in the maximum angle of insolation for a location at 42°N latitude is shown in Figure 6-4. For each degree of latitude a location is away from the position of 90° (perpendicular) insolation, the angle of insolation is one degree less. (See Table 6-1 on page 71.)

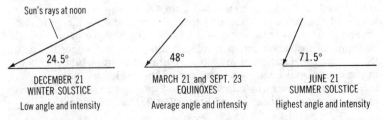

Figure 6-4. Maximum angle of insolation (at noon) at 42° north latitude at different seasons.

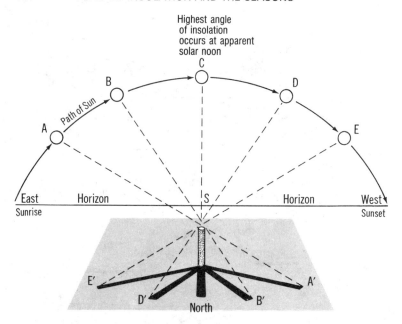

Figure 6-5. The shadow of a vertical post indicates how the angle of insolation varies during the day. The higher the angle of insolation, the shorter the shadow and the greater the intensity. Maximum angle and intensity occur at apparent solar noon.

INTENSITY OF INSOLATION IN RELATION TO TIME OF DAY. The angle of insolation is the same as the altitude of the sun. On any given day, the altitude of the sun varies from zero at sunrise, to a maximum at apparent solar noon, and back to zero again at sunset. The angle of insolation, and hence the intensity of insolation, also vary during the day in the same way, as shown by the changing length of the shadow in Figure 6-5.

DURATION OF INSOLATION. **Duration of insolation** is the length of time that insolation is received each day. That is, it is the number of hours that the sun is visible each day. The duration of insolation varies with latitude and with the seasons. Figure 4-6 on page 34 shows how the length of the sun's path through the sky varies during the year. The duration of insolation is proportional to the length of the path. There is one hour of insolation for each 15° of path. Thus, the longer the path, the greater the duration of insolation. Duration of insolation is greatest in the continental United States at the summer solstice (June 21), least at the winter solstice (December 21), and an average length, 12 hours, at the equinoxes (March 21 and September 23). See Table 6-1.

TEMPERATURES IN RELATION TO INSOLATION. The surface of the earth continuously radiates energy, mostly in the infrared wave-

TABLE 6-1. ANGLE AND DURATION OF INSOLATION

Latitude	SUMMER SOLSTICE JUNE 21		EQUINOXES MARCH 21 SEPTEMBER 23		WINTER SOLSTICE DECEMBER 21	
	Angle of Insolation at 12 Noon	Duration of Insolation	Angle of Insolation at 12 Noon	Duration of Insolation	Angle of Insolation at 12 Noon	Duration of Insolation
90°N	23½°	24 Hours	0°	12 Hours	—	0 Hours
80°N	33⅓°	24	10°	12	—	0
70°N	43½°	24	20°	12	—	0
66½°N	47°	24	23½°	12	0°	0
60°N	53½°	18½	30°	12	6½°	5½
50°N	63½°	16¼	40°	12	16½°	7¾
40°N	73½°	15	50°	12	26½°	9
30°N	83½°	14	60°	12	36½°	10
23½°N	90°	13½	66½°	12	43°	10½
20°N	86½°	13¼	70°	12	46½°	10¾
10°N	76½°	12½	80°	12	56½°	11½
0°	66½°	12	90°	12	66½°	12
10°S	56½°	11½	80°	12	76½°	12½
20°S	46½°	10¾	70°	12	86½°	13¼
23½°S	43°	10½	66½°	12	90°	13½
30°S	36½°	10	60°	12	83½°	14
40°S	26½°	9	50°	12	73½°	15
50°S	16½°	7¾	40°	12	63½°	16¼
60°S	6½°	5½	30°	12	53½°	18½
66½°S	0°	0	23½°	12	47°	24
70°S	—	0	20°	12	43½°	24
80°S	—	0	10°	12	33½°	24
90°S	—	0	0°	12	23½°	24

lengths. The rate of radiation depends on the temperature, being greater when the temperature of the surface is higher, and less when it is lower. The temperature of the earth's surface at any particular location varies through the day, and the average daily temperature varies through the year. This variation depends on the balance between the energy being gained from insolation and the energy being lost by terrestrial radiation. When the energy is being gained at a greater rate than it is being lost, the temperature rises. When energy is being lost faster than it is being gained, the temperature falls. Temperatures are generally higher when the intensity of insolation is greater; they are also higher when the duration of insolation is longer.

TIMES OF MAXIMUM AND MINIMUM TEMPERATURES.

1. **During the year.** It might be thought at first that the time of maximum temperature would occur at the time of maximum insolation, and the minimum temperature at the time of minimum insolation. This

is not the case, however. The times of maximum and minimum temperature occur somewhat later than the times of maximum and minimum insolation. In latitudes north of 23½°N, for example, the maximum intensity of insolation and the maximum duration of insolation occur on June 21, but the time of highest daily temperatures occurs toward the end of July or early in August. (See Figure 6-6.)

The reason for this is that as the intensity and duration of insolation increase during the spring, each day the surface receives more energy of insolation than it loses by radiation. Consequently, each day the average temperature goes up slightly. On June 21, more energy is gained than is lost; the temperature is still rising as a result. On June 22, a little less energy is received than on June 21. However, the amount received is still more than the amount lost. Therefore the temperature continues to go up. This continues until the rate of incoming energy finally drops below the rate of energy loss by radiation. At that time (usually in late July or early August) the surface temperature begins to drop.

Similarly, the earth north of 23½°N is not at its coolest during the time of minimum angle and duration of insolation on December 21. This occurs because the area continues to lose more energy than is gained by insolation until some time in January or February, when the region reaches its coolest temperature.

2. During the day. The hottest part of an average day is some time in midafternoon, not at apparent solar noon when insolation is greatest (see Figure 6-7). The coolest part of an average day is slightly after sunrise because the earth continues to lose heat during nighttime until after insolation begins at sunrise.

RADIATIVE BALANCE. If an object is in **radiative balance,** it gains the same amount of energy that it gives off, and average temperatures remain the same. Points X and Y in Figure 6-6 indicate the times when a given place on earth is in radiative balance. The graph (Figure 6-8)

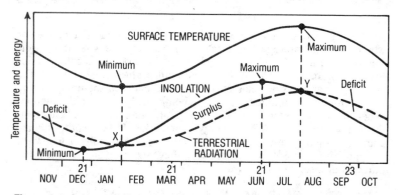

Figure 6-6. Average daily temperatures, insolation, and terrestrial radiation for mid-latitudes of the Northern Hemisphere during a year.

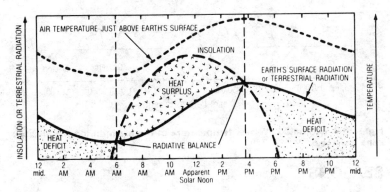

Figure 6-7. Typical variation in air temperature just above surface, insolation, and terrestrial radiation in the course of one day.

can be used to illustrate average temperatures for the eastern United States, which we will assume reflect average temperature changes of the earth. This graph illustrates the following points:

1. Annual measurements of temperature indicate (by changing) that the earth is not in radiative balance on a yearly basis.

2. Intermediate measurements (decades) indicate little change in average decade temperatures; thus the earth is in radiative balance over periods of decades.

The graph in Figure 6-9 on page 74 illustrates that the earth is not in radiative balance over very large periods of time. At different times during the earth's history, the earth has cooled off and glaciers have covered large portions of the earth's surface.

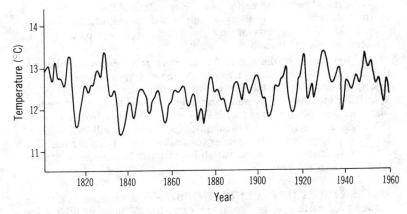

Figure 6-8. Estimated annual temperatures of the eastern seaboard of the United States, centered on Philadelphia, for a period of several decades.

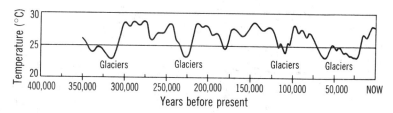

Figure 6-9. Estimated average temperature of the earth during the Pleistocene Ice Age.

SEASONS

The **seasons** are the four divisions of the year that are characterized by different types of weather conditions. For most of the earth, the major difference in the seasons is difference in temperature.

DIRECT CAUSE OF THE SEASONS. The seasonal changes in temperature and other weather conditions are the *direct* result of the cyclic variations in intensity and duration of insolation that occur during the year. These variations were described on pages 68–70.

INDIRECT CAUSES OF THE SEASONS. The variations in insolation that cause the seasons are themselves the result of factors that may be called the *indirect* causes of the seasons. These indirect causes are the following:

1. **Tilt of the earth's axis.** As Figure 6-10 (page 75) shows, the earth's axis is inclined at an angle of 23½° with respect to a line perpendicular to the plane of its orbit.

2. **Parallelism of the earth's axis.** Regardless of the position of the earth in its orbit, its axis always points in the same direction in space. The direction of the axis at any given time is always parallel to its direction at any other time.

3. **Revolution of the earth.** As the earth revolves around the sun, the direction of the axis with respect to the sun varies, because of the first two factors described above. For example, on June 21 the North Pole is inclined *toward* the sun at an angle of 23½° from the perpendicular. On December 21, the North Pole is inclined *away* from the sun at this angle. On March 21 and September 23, the axis is still inclined 23½° from the perpendicular, but *neither* toward *nor* away from the sun. This cycle of variations causes the variations in angle and duration of insolation through the year.

SMALL EFFECT OF ELLIPTICAL ORBIT. The variation in distance between the earth and the sun as the earth travels along its orbit is too small to have a significant effect on the seasons. For example, winter in the Northern Hemisphere occurs at a time when the earth is actually nearest the sun.

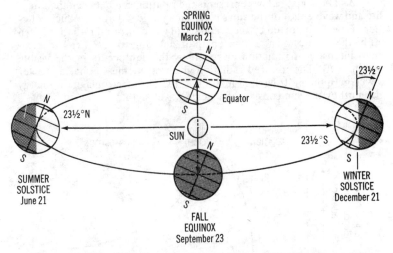

Figure 6-10. Causes of the seasons. The tilt of the earth's axis and parallelism of the axis result in varying angles and duration of insolation through the year.

VOCABULARY	
insolation	perpendicular insolation (direct or
aerosols	vertical rays)
terrestrial radiation	solstice
angle of insolation	duration of insolation
greenhouse effect	radiative balance
intensity of insolation	seasons
equinox	

QUESTIONS ON TOPIC VI—INSOLATION AND THE SEASONS
Questions in Recent Regents Exams (end of book)

June 1995:   10, 11, 16, 18, 26, 58
June 1996:   9–11, 14, 15, 17, 66–70
June 1997:   14–16, 67, 78–80
June 1998:   5, 11, 15, 16, 67, 69, 70

Questions from Earlier Regents Exams

1. Electromagnetic energy that reaches the earth from the sun is called   (1) insolation   (2) conduction   (3) specific heat   (4) terrestrial radiation

2. The radiation that passes through the atmosphere and reaches the earth's surface has the greatest intensity in the form of   (1) visible-light radiation   (2) infrared radiation   (3) ultraviolet radiation   (4) radio-wave radiation

3. The graph below represents the relationship between the intensity and wavelength of the sun's electromagnetic radiation. Which statement is best supported by the graph?   (1) The infrared radiation given off by the sun occurs at a wavelength of 2000 angstroms.   (2) The maximum intensity of radiation given off by the sun occurs in the visible region.   (3) The infrared radiation given off by the sun has a shorter wavelength than ultraviolet radiation.   (4) The electromagnetic energy given off by the sun consists of a single wavelength.

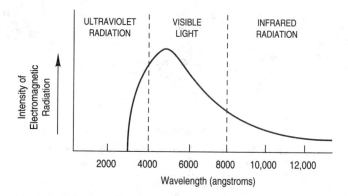

4. Water vapor and carbon dioxide in the earth's atmosphere are good absorbers of   (1) visible radiation   (2) infrared radiation   (3) ultraviolet radiation   (4) x rays

Base your answers to questions 5 and 6 on the diagram below, which represents a field showing the amount of insolation received at the earth's surface in calories per square centimeter per minute on a clear day in the morning.

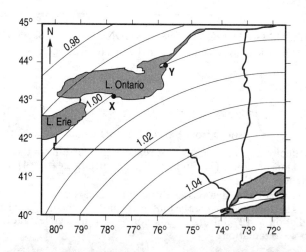

5. If the insolation value at the outer edge of the earth's atmosphere over New York State equals 1.85 calories per square centimeter per minute, which best explains why the values are lower at the surface in New York State?   (1) The angle of insolation is greater in New York State than at the equator.   (2) Insolation decreases with decreased latitude.   (3) The energy is used to evaporate water.   (4) Insolation is absorbed and reflected by the atmosphere.

6. Which change in atmospheric conditions within the dashed circle could cause the different pattern of isolines as indicated in that area in the accompanying diagram?

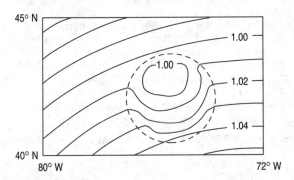

(1) decreasing dew point temperatures   (2) decreasing velocity of a north wind   (3) increasing cloud cover   (4) increasing air pressure

7. Which form of electromagnetic energy is radiated from the earth's surface with the greatest intensity?   (1) x rays   (2) infrared rays   (3) ultraviolet rays   (4) visible light rays

8. Electromagnetic energy that is being given off by the surface of the earth is called   (1) convection   (2) insolation   (3) specific heat   (4) terrestrial radiation

9. Which statement best explains why, at high latitudes, reflectivity of insolation is greater in winter than in summer?   (1) The North Pole is tilted toward the sun in winter.   (2) Snow and ice reflect almost all insolation.   (3) The colder air holds much more moisture.   (4) Dust settles quickly in cold air.

10. Adding more carbon dioxide to the atmosphere increases the amount of   (1) radiant energy reflected by the earth   (2) radiation from the sun absorbed by the oceans   (3) radiation from the earth absorbed by the atmosphere   (4) ultraviolet rays striking the earth

11. Some scientists predict that the increase in atmospheric carbon dioxide will cause a worldwide increase in temperature. Which could result from this increase in temperature?   (1) Continental drift will increase.   (2) Isotherms will shift toward the equator.   (3) Additional landmasses will form.   (4) Ice caps at the earth's poles will melt.

12. Which model best represents how a greenhouse remains warm as a result of insolation from the sun?

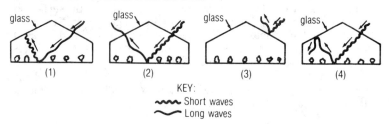

KEY:
∿∿∿ Short waves
⌒ Long waves

13. Compared to polar areas, why are equatorial areas of equal size heated much more intensely by the sun?   (1) The sun's rays are more nearly perpendicular at the equator than at the poles.   (2) The equatorial areas contain more water than the polar areas do.   (3) More hours of daylight occur at the equator than at the poles.   (4) The equatorial areas are nearer to the sun than the polar areas are.

14. In which diagram does the incoming solar radiation reaching the earth's surface have the greatest intensity?

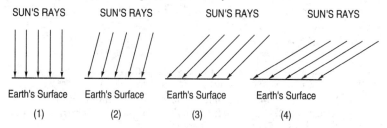

15. The most logical conclusion that can be made from the relationship between the altitude of the sun throughout the day and the amount of insolation is that, as the sun's altitude   (1) increases, the insolation increases   (2) increases, the insolation decreases   (3) decreases, the insolation increases   (4) decreases, the insolation remains the same

16. In New York State at 3 P.M. on September 21, the vertical pole shown in the diagram casts a shadow. Which line best approximates the position of that shadow?

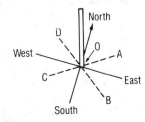

(1) *OA*   (2) *OB*   (3) *OC*   (4) *OD*

17. At the time of the fall equinox, the number of hours of daylight in New York City is generally about   (1) nine   (2) twelve   (3) fifteen   (4) eighteen

Base your answers to questions 18 through 21 on the diagrams below. The diagrams represent models of the apparent path of the sun across the sky for observers at four different locations A through D on the earth's surface.

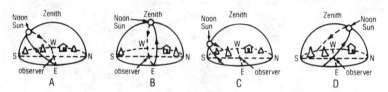

18. At location A, on which side of an observer would his shadow fall at local noon?   (1) south side   (2) north side   (3) east side   (4) west side

19. If the model of location B represents the apparent path of the sun observed at the equator, what is the date at location B?   (1) March 21   (2) June 21   (3) October 21   (4) December 21

20. If the model of location D represents the apparent path of the sun on December 21, where is location D?   (1) the North Pole   (2) 45° N latitude   (3) the equator   (4) 45° S latitude

21. At location B three months later, how would the altitude of the noon sun compare to its present altitude?   (1) The altitude would be less than shown.   (2) The altitude would be greater than shown.   (3) The altitude would be the same as shown.

22. In New York State, summer is warmer than winter because in summer, New York State has   (1) more hours of daylight and is closer to the sun   (2) more hours of daylight and receives more direct insolation   (3) fewer hours of daylight but is closer to the sun   (4) fewer hours of daylight but receives more direct insolation

23. The diagram below represents a model of the sun's apparent path across the sky in New York State for selected dates.

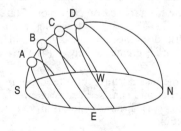

For which path would the duration of insolation be greatest?   (1) A   (2) B   (3) C   (4) D

Base your answers to questions 24 through 28 on the diagram below. The diagram represents the earth at a specific time in its orbit. The dashed lines indicate radiation from the sun. Points *A* through *H* are locations on the earth's surface.

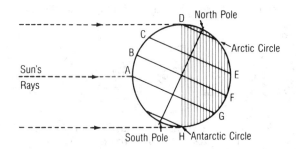

24. Which line represents the equator?   (1) *AG*   (2) *BF*   (3) *CE*   (4) *DH*

25. What is the season in the Northern Hemisphere when the earth is in the position shown in the diagram?   (1) spring   (2) summer   (3) fall   (4) winter

26. When the sun is in the position shown in the diagram, how many hours of daylight would occur at the North Pole during one complete rotation?   (1) 0   (2) 8   (3) 12   (4) 24

27. In which direction would a person located at position *H* have to look to see the sun at the time shown in the diagram?   (1) north   (2) east   (3) south   (4) west

28. Six months after the date indicated by the diagram, which point would receive the sun's vertical rays at noon?   (1) *A*   (2) *B*   (3) *C*   (4) *D*

29. Which two factors determine the number of hours of daylight at a particular location?   (1) longitude and season   (2) longitude and the earth's average diameter   (3) latitude and season   (4) latitude and the earth's average diameter

30. What is the primary reason New York State is warmer in July than in February?   (1) The earth is traveling faster in its orbit in February.   (2) The altitude of the noon sun is greater in February.   (3) The insolation in New York is greater in July.   (4) The earth is closer to the sun in July.

31. What is the usual cause of the drop in temperature that occurs between sunset and sunrise at most New York State locations?   (1) strong winds   (2) ground radiation   (3) cloud formation   (4) heavy precipitation

32. In New York State, the maximum total daily insolation occurs during June. Which statement best explains why the maximum annual temperature is usually observed about a month later, in July?   (1) The earth is closer to the sun in June than it is in July.   (2) The earth is farther from the sun in June than it is in July.   (3) New York State loses more energy than it receives during most of July.   (4) New York State receives more energy than it loses during most of July.

33. Between the years 1850 and 1900, records indicate that the earth's mean surface temperatures showed little variation. This would support the inference that   (1) the earth was in radiative balance   (2) another ice age was approaching   (3) more energy was coming in than was going out from the earth   (4) the sun was emitting more energy

34. Which is the main cause of seasons on the earth?   (1) difference in the earth-sun distance at perihelion and aphelion   (2) inclination of the earth's axis of rotation to the plane of its orbit   (3) change of inclination of the earth's axis of rotation in relation to the plane of its orbit   (4) increase in velocity of the earth as it approaches the sun

35. Which position in the diagram below best represents the earth on the first day of summer in the Northern Hemisphere?

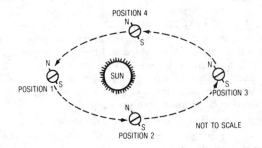

(1) 1   (2) 2   (3) 3   (4) 4

36. Which change would occur if the earth's axis were inclined at an angle of 33½° instead of 23½°?   (1) The equator would receive fewer hours of daylight on June 21.   (2) The sun's direct ray would move over a larger area of the earth's surface.   (3) The celestial equator would be farther from the zenith.

37. Which is *not* a cause of our change of seasons?   (1) revolution of the earth   (2) inclination of the earth's axis   (3) variation of the distance to the sun   (4) parallelism of the earth's axis

Variations in insolation (discussed in Topic VI) cause heat energy to be unevenly distributed in the atmosphere. This heat energy tends to move toward a condition of more uniform distribution. That movement of heat energy results in the constant changes in the atmosphere that are a major cause of *weather*.

Weather is the state or condition of the variables of the atmosphere at a location for any given short period of time. **Atmospheric variables** include temperature, pressure, wind, and moisture. Most of the weather changes occur in the troposphere, the part of the atmosphere immediately above the earth's surface. (For the location of the divisions of the atmosphere, see Figure 3-6, page 16.)

ATMOSPHERIC TEMPERATURE

Temperature data is often shown on maps by the use of isolines called **isotherms.**

HEATING OF THE ATMOSPHERE.   The sun is the original source of heat for the atmosphere. Generally, the more insolation at a location, the warmer will be the earth's surface and the atmosphere above it. The atmosphere acquires much of its heat directly from the earth's surface, but it also gains energy in other ways. The many ways in which the atmosphere is heated (gains energy) include the following:

1. Conduction by contact with the earth's surface.

2. Direct absorption of insolation from the sun.

3. Absorption of radiations from the earth's surface. (Much of the heat absorption by the atmosphere is due to the presence of water vapor and carbon dioxide; the larger the amounts of water vapor and carbon dioxide, the more heat is absorbed by the atmosphere.)

4. **Condensation** (change of water vapor to liquid water) and **sublimation** (change of water vapor directly to ice) release large amounts of latent heat, directly heating the atmosphere.

5. The Coriolis effect, resulting from the rotation of the earth, causes a **frictional drag** at the interface of the atmosphere and the earth's surface; this friction produces heat, some of which is added to the atmosphere.

TRANSFER OF HEAT IN THE ATMOSPHERE.   Differences in air density cause differences in air pressure. Air pressure differences in turn cause air to move parallel to the earth's surface *(wind)* and in vertical circular patterns called *convection cells* or *currents*. These air movements transfer heat energy within the atmosphere.

ADIABATIC HEATING AND COOLING. When a gas expands, its temperature decreases; when a gas is compressed, its temperature increases. This automatic change in the temperature of a gas due to expansion or compression is called an **adiabatic temperature change.** Thus when air rises in the atmosphere, it expands and its temperature decreases adiabatically; similarly, when it descends, it is compressed and its temperature increases. In the troposphere, temperatures generally decrease with altitude as a result of adiabatic processes. (See *Selected Properties of Earth's Atmosphere* in the *Earth Science Reference Tables.*)

ATMOSPHERIC PRESSURE AND DENSITY

The more dense the atmosphere, the greater the weight of a given volume and therefore the greater the air pressure exerted by it; thus air pressure and density are directly related. **Atmospheric pressure,** also called **barometric pressure** and **air pressure,** is the pressure due to the weight of the overlying atmosphere pushing down on any given area. Standard air pressure (one atmosphere) at sea level is 29.92 inches of mercury or 1013.2 millibars. Barometric pressure in millibars can be converted to inches of mercury using the scale in the *Earth Science Reference Tables.* Air pressure is measured by an instrument called a barometer. Air pressure is often shown on weather maps by the use of isolines called **isobars.**

EFFECT OF TEMPERATURE ON PRESSURE. Changes in the temperature of the air cause changes in air pressure. As the temperature of air increases, the air expands, so that its density and pressure decrease. Decreasing temperature has the reverse effect.

EFFECT OF MOISTURE ON PRESSURE. The greater the amount of water vapor in the air (called **moisture** content, absolute humidity, and vapor pressure), the lower the air density and pressure. The reason for this is that each water molecule in the atmosphere *replaces* a molecule of air, usually oxygen or nitrogen. Since a water molecule weighs less than either an oxygen or a nitrogen molecule, the greater the amount of water vapor in the air, the less dense the air as a whole becomes.

EFFECT OF ALTITUDE ON PRESSURE. As altitude increases, atmospheric density and pressure decrease. (See *Selected Properties of Earth's Atmosphere* in the *Earth Science Reference Tables.*)

To summarize, as either altitude, temperature, or moisture content in the atmosphere increases, air density and air pressure decrease.

WIND

Horizontal movement of air parallel to the earth's surface is called **wind.** Wind is a vector field, because it requires a magnitude and direction to describe it.

WIND SPEED.   Winds are caused by differences in air pressure. The difference in air pressure for a specific distance is called the **pressure gradient.** The closer together the isobars on a weather map are, the greater (steeper) the pressure gradient. The greater the pressure gradient is, the greater the wind speed (see Figure 7-1).

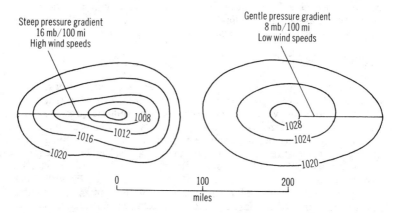

Figure 7-1. Pressure gradient as indicated by closeness of isobars on a weather map. The greater the pressure gradient, the greater the wind speed.

WIND DIRECTION.   Air (wind) moves from areas of high pressure to areas of low pressure. However, the Coriolis effect (caused by the earth's rotation) modifies that pattern of movement, deflecting winds to the right in the Northern Hemisphere and to the left in the Southern Hemisphere. Figure 7-2 illustrates how the wind directions are modified by the Coriolis effect.

A wind is named for the direction from which it comes; for example, a wind blowing *from* the south *toward* the north is called a *south* wind.

Figure 7-2. Deflection of winds by the Coriolis effect. The dashed arrow in each case shows the direction the wind would flow if there were no Coriolis effect. The solid arrow shows the actual path of the wind. If you face with the wind, it is always deflected to the right in the Northern Hemisphere and to the left in the Southern Hemisphere.

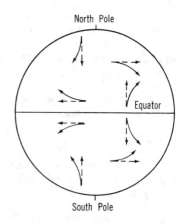

CIRCULATION OF THE ATMOSPHERE

CONVECTION CELLS. The unequal distribution of insolation on the earth results in unequal heating and differences in air pressure. Cooler air, being denser, sinks toward the earth under the influence of gravity, causing the less dense warmer air to rise. The result is a series of convection cells around the earth at various latitudes, as shown in Figure 7-3. As indicated by the solid arrows, there are upward currents in the vicinity of 0° latitude (the equator) and 60° north and south latitudes. Downward currents exist near 30° and 90° north and south latitudes. Regions where air comes together to form vertical currents are regions of **convergence** (labeled "C"). Regions where air spreads out from the vertical currents are regions of **divergence** (labeled "D").

PLANETARY WINDS. At the earth's surface, winds flow horizontally away from regions of divergence and toward regions of convergence. Because of the Coriolis effect, these winds are deflected to the right in the Northern Hemisphere and to the left in the Southern Hemisphere, as indicated by the dashed arrows in Figure 7-3. The result is a series of **planetary wind belts** within which the winds move generally in a specific direction much of the time. The *jet streams* are wavy bands of high-speed winds located near the top of the troposphere.

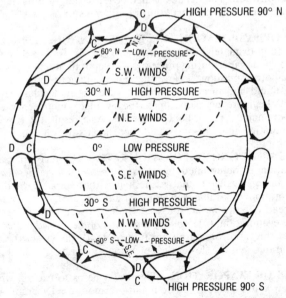

Figure 7-3. Planetary wind and pressure belts in the troposphere. The drawing shows the locations of the belts near the time of an equinox. The locations shift somewhat with the changing latitude of the sun's vertical ray. In the Northern Hemisphere the belts shift northward in summer and southward in winter. (Also see *Planetary Winds and Moisture Belts in the Troposphere* in the *Earth Science Reference Tables*.)

WEATHER MOVEMENT. Much of the continental United States is affected by planetary winds that blow from the southwest. Therefore, weather changes in the United States move generally from a southwesterly direction to a northeasterly direction (see Figure 8-11).

SURFACE OCEAN CURRENTS. The surface ocean currents are caused by wind blowing over the oceans and transferring energy to the water. The direction of these currents is affected by the direction of the planetary winds. Compare *Surface Ocean Currents* with *Planetary Winds and Moisture Belts in the Troposphere* in the *Earth Science Reference Tables*.

ATMOSPHERIC MOISTURE

Atmospheric moisture exists in three states of matter: liquid, solid, and gas. Gaseous water in the atmosphere is called **water vapor**.

SOURCES OF ATMOSPHERIC MOISTURE. Moisture in the form of water vapor enters the atmosphere by *evaporation*, *sublimation*, and *transpiration*. **Evaporation** is the process by which a liquid changes to a gas; **transpiration** is the process by which plants release water vapor as part of their life functions. Collectively, evaporation and transpiration are called **evapotranspiration**. The oceans, which cover about 70% of the earth's surface, are the source of most atmospheric moisture.

ENERGY OF EVAPORATION AND TRANSPIRATION. Large amounts of energy (approximately 540 calories/gram) are required to change liquid water into water vapor during processes of evaporation and transpiration. When evapotranspiration occurs, the more energetic water molecules leave the liquid and form water vapor. Since the molecules leaving the liquid are the more energetic ones, the average kinetic energy of the molecules remaining in the liquid decreases. As a result, the temperature of an evaporating liquid is somewhat lower than its surroundings.

FACTORS AFFECTING EVAPORATION RATE. The net evaporation rate at a location is determined by (1) *the amount of energy available*— the more energy available, that is, the higher the temperature, the faster the evaporation of available water; (2) *surface area of the water*—the more spread out the water, the greater the air-water interface and the faster the evaporation; (3) *degree of saturation of the air with water vapor*— the higher the vapor pressure (moisture content), the closer the air is to being saturated and the slower the net evaporation rate.

PROCESS OF EVAPORATION. Figure 7-4 helps to explain what happens when water—or any other liquid—evaporates. Diagram A shows a closed container of air into which some water has just been added. We assume there are no water vapor molecules in the air at this moment. However, as soon as water is placed in the container, some of its more energetic molecules begin to escape into the air above and mingle with the other gas molecules that are present. Diagram B shows the situation a short time later. There are now some water molecules in the air, represented by the four molecules shown. We see two more molecules in the process of leaving the water. However, one of the other four is about to enter the water, so that there will be five molecules in the air after this happens. Diagram C shows the situation some time later. There are now eight water molecules in the air. Again, two more water molecules are about to leave the water. But now *two* of the eight molecules in the air are about to return to the water. The net result wil be eight molecules in the air—the same number as before. We have a condition of *dynamic equilibrium*, in which the rate at which water is entering the gaseous state (evaporation) equals the rate at which water vapor is returning to the liquid state (condensation).

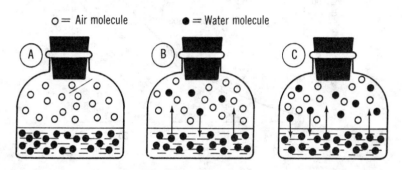

Figure 7-4. The saturation of the air over an evaporating liquid. At A, no evaporation has occurred. At B, evaporation is proceeding faster than condensation. At C, equilibrium has been reached. The air is **saturated** with vapor, and evaporation and condensation are occurring at the same rate. The pressure of the vapor under these conditions is called the **saturation vapor pressure**. The higher the temperature, the greater the saturation vapor pressure.

SATURATION VAPOR PRESSURE.   In the equilibrium condition described above, the air contains the maximum amount of water vapor it can hold (represented by eight molecules). This conditions is called **saturation**. Water vapor, being a gas, exerts a pressure, which is called **vapor pressure**. When the air is saturated with water vapor (at capacity), the vapor pressure is called the **saturation vapor pressure**.

ABSOLUTE HUMIDITY.   Humidity, like moisture, is a general term that refers to the water vapor content of the atmosphere. The amount of water vapor in each unit volume of air is called the **absolute humidity**. There is a definite vapor pressure corresponding to each value of absolute humidity. The vapor pressure increases in direct proportion to the absolute humidity, so that vapor pressure can be used as a measure of absolute humidity.

ABSOLUTE HUMIDITY AND TEMPERATURE.   The maximum absolute humidity and saturation vapor pressure (and capacity) increase rapidly with an increase in temperature, as shown in the graph in Figure 7-5. This means that the moisture capacity of the air increases with an increase in temperature.

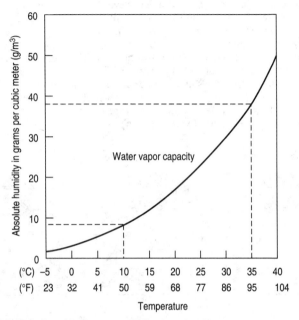

Figure 7-5. **Relationship between air's capacity for water vapor and temperature.** Note that air with a temperature of 35°C can hold about four times more water vapor than can air at 10°C.

RELATIVE HUMIDITY.   **Relative humidity** is the ratio of the amount of water vapor in the air to the maximum amount it can hold (its **moisture capacity**). It is also the ratio of the absolute humidity/maximum absolute humidity (capacity). The closer the absolute humidity is to the maximum absolute humidity, the higher the relative humidity. (Be sure to understand that relative humidity is a percentage of saturation, *not* a percentage of the volume of air present.)

RELATIVE HUMIDITY AND TEMPERATURE.   At any given time and place, the air has a certain amount of water vapor (humidity), with a corresponding absolute humidity. If the temperature of the air changes, but the amount of water vapor (that is, the absolute humidity) remains the same, the relative humidity will change. For example, if the temperature increases, the relative humidity will decrease. This occurs because the capacity increases with the increase in temperature, while the absolute humidity remains the same. On the other hand, if the temperature decreases while the absolute humidity remains the same, the relative humidity will increase.

RELATIVE HUMIDITY AND ABSOLUTE HUMIDITY.   If the temperature of the air remains constant, but more water vapor is added to it (for example, by evapotranspiration), the absolute humidity, the vapor pressure, and the relative humidity will all increase.

DEW POINT TEMPERATURE.   If the temperature of the air decreases while the absolute humidity remains the same, the temperature will eventually reach a point at which the absolute humidity equals the maximum absolute humidity. At this temperature, the relative humidity will be 100%, that is, the air will be saturated. This temperature is called the **dew point temperature**, or simply the **dew point**. Any further drop in temperature will result in condensation or sublimation of water, since there would then be more water vapor in the air than it can hold.

DEW POINT AND ABSOLUTE HUMIDITY.   The dew point temperature depends only on the absolute humidity or vapor pressure of the air, and not on the relative humidity. As the vapor pressure or the amount of water vapor in the air increases, the dew point rises.

RELATIVE HUMIDITY.   It is difficult to measure absolute humidity or determine the air's capacity to hold water vapor directly. Relative humidity is therefore measured by indirect methods. One such method uses an instrument called a *sling psychrometer*, as shown in Figure 7-6. This instrument contains an ordinary thermometer called the **dry-bulb thermometer,** and another thermometer with a wick around its bulb, called the **wet-bulb thermometer.** When the wick is moistened and the thermometers are whirled in the air, the temperature of the wet-bulb drops because of the cooling effect of the evaporation of the water. The amount of cooling depends on the rate of evaporation, and is therefore related to the relative humidity.

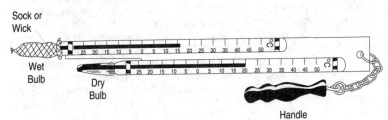

Figure 7-6. Sling psychrometer. The difference between the two temperature readings is used to find the relative humidity.

MEASURING DEWPOINT TEMPERATURE.   To determine the dewpoint temperature, you need the dry-bulb and wet-bulb readings from a sling psychrometer and the *Dewpoint Temperatures* chart in the *Earth Science Reference Tables*.

1. Locate the dry-bulb reading on the left-hand side of the chart. (Note that the dry-bulb reading is the same as a regular temperature reading.)

2. Subtract the wet-bulb reading from the dry-bulb reading.

3. Locate the difference between the wet-bulb and dry-bulb readings across the top of the chart.

4. Follow the horizontal row for the dry-bulb reading to the right until it meets the vertical column running down from the difference between the wet-bulb and dry-bulb readings. This number is the dewpoint temperature.

As an example, suppose the dry-bulb reading is 8°C and the wet-bulb reading is 6°C. The difference is 2°C. Reading across from 8°C and down from 2°C, you find a dewpoint temperature of 3°C.

MEASURING RELATIVE HUMIDITY (%).   Determining the percent relative humidity is similar to determining the dewpoint temperature.

1. Locate the dry-bulb reading on the left-hand side of the *Relative Humidity (%)* chart in the *Earth Science Reference Tables*.

2. Subtract the wet-bulb reading from the dry-bulb reading.

3. Locate the difference between the wet-bulb and dry-bulb readings across the top of the chart.

4. Follow the horizontal row for the dry-bulb reading to the right until it meets the vertical column running down from the difference between the wet-bulb and dry-bulb readings. This number is the relative humidity.

As an example, suppose the dry-bulb reading is 4°C and the wet-bulb reading is −2°C. The difference is 6°C. Reading across from 4°C and down from 6°C, you find a relative humidity of 14%.

CONDENSATION, SUBLIMATION, AND CLOUDS.   If the temperature of the air cools (often adiabatically) below the dewpoint, the water vapor will usually condense or sublime to a solid, changing to microscopic liquid water droplets or ice crystals. In the atmosphere, if the temperature is above 0°C, condensation produces water droplets which as a group appear as a *cloud*. Some clouds are composed of ice crystals, which form when water vapor sublimates to a solid. A **cloud** is therefore a collection of liquid water droplets and/or ice crystals suspended in the atmosphere. This condensation or sublimation to form clouds releases tremendous amounts of energy to the troposphere. At the earth's surface, condensation produces dew and sublimation produces frost.

Since temperature and dewpoint temperature both decrease with elevation in the troposphere, the altitude where temperature and dewpoint are the same is the level where clouds form by condensation or sublimation. This is often illustrated by a lapse rate (change rate) chart like the *Lapse Rate* chart in the *Earth Science Reference Tables*.

Refer to the *Lapse Rate* chart. On this chart, the dashed lines represent the decrease in dewpoint with increasing altitude (2C°/km). The solid diagonal lines represent the decrease in temperature with increasing altitude (10C°/km) above the earth's surface. If you know the temperature and dewpoint at the earth's surface, you can determine the altitude of the condensation (sublimation) level and therefore the altitude of the bottom of the clouds above a location.

Suppose the surface temperature is 20°C and the surface dewpoint is 10°C. Follow these steps to find the level at which clouds will form.

1. Find the 20°C mark and the 10°C mark along the bottom edge of the chart.

2. From the 20°C mark, follow the solid diagonal line upward and to the left.

3. From the 10°C mark, follow the dashed line upward.

4. The place where the two lines intersect represents the altitude at which clouds will form, in this case, about 1,200 km above the surface.

CONDENSATION SURFACE.   Besides needing saturated air, condensation requires a surface upon which the vapor can condense. This is called the **condensation surface**. In the formation of clouds, the condensation surfaces are aerosols such as dust and volcanic ash.

PRECIPITATION.   **Precipitation** is the falling of liquid or solid water from clouds toward the surface of the earth. The forms in which precipitation occurs are rain, snow, sleet, drizzle, freezing rain, and hail. For precipitation to occur, the ice crystals or water droplets in clouds must become big enough so that they will fall.

ATMOSPHERIC TRANSPARENCY AND PRECIPITATION.   The more pollutants added to the atmosphere by the activities of nature and people, the more aerosols are present. The more aerosols in the air, the less transparent the atmosphere is to insolation. "Less transparent" means that more energy from the sun is absorbed or reflected by the atmosphere, resulting in less insolation reaching the earth's surface. When the atmosphere has a high enough aerosol content so that distant images are blurred and a cloudless sky does not appear clear blue, the condition is called *haze*. Condensation in cloud formation incorporates some of the aerosols, and these aerosols are removed from the atmosphere during precipitation. The falling liquid or solid water also collects other aerosols on the way down, thus lowering air pollution levels and cleaning the atmosphere.

AIR MASSES AND FRONTS

Much of the weather of the continental United States is the result of the invasion of *air masses* and their interactions. An **air mass** is a large body of air in the troposphere with similar characteristics of pressure, moisture, and temperature.

SOURCE REGIONS.   An air mass forms when a large mass of air stagnates over a part of the earth's surface for a period of time and thus acquires some of the characteristics of that surface. The geographic regions in which air masses are formed are called **source regions.** The source regions for the air masses that invade the United States are shown in Figure 7-7.

If the source region is at a high latitude, the air mass will have a low temperature; if the source region is at a low latitude, the air mass will have a high temperature. If the source region is land, the air mass will be dry; if the source region is water, the air mass will be moist. Air masses are described as arctic (A), polar (P), or tropical (T), depending on whether they originate at high or low latitudes. Air masses are also described as either continental (c) or maritime (m), depending on whether they originate over land or water. These air mass symbols are found in the *Earth Science Reference Tables.* When an air mass remains stationary over its source region for a period of time, it tends to acquire relatively uniform temperature and moisture conditions at any one altitude within it.

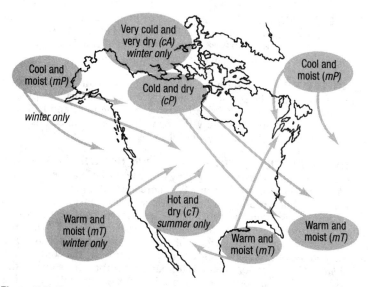

Figure 7-7. Source regions and tracks of air masses that affect the weather of the continental United States.

LOWS AND HIGHS.   Portions of the troposphere are divided into two types according to pressure and direction of circulation of winds. **LOWS,** also called **cyclones,** have low pressure. The pressure in a LOW is lowest at its center, and thus winds blow toward the center. The winds are deflected by the Coriolis effect, so that in the Northern Hemisphere a counterclockwise circulation occurs within LOWS, as shown in Figure 7-8. **HIGHS,** or **anticyclones,** have high pressure. The pressure in a HIGH is greatest at its center, and thus winds blow out from the center. The general circulation around HIGHS is clockwise in the Northern Hemisphere.

FRONTS. Where two air masses of different characteristics meet, an interface called a **front** develops (see Figure 7-9 on page 94). A **cold front** is the interface between an advancing cold air mass and a warmer air mass, where the underlying cold air pushes forward like a wedge. A **warm front** is a front between an advancing warm air mass and a retreating wedge of a cooler air mass; because the cool air is heavier, the warm air mass is forced to rise as it advances. An **occluded front** is the interface between opposing wedges of cold air masses formed when a cold front overtakes a warm front, lifting the warm air mass off the ground. Occluded fronts are important because they are associated with the formation of mid-latitude cyclones (LOWS), as shown in Figure 7-8. When two adjacent air masses of different characteristics remain in the same positions, they form a **stationary front**.

At fronts between air masses of different temperatures the warmer air, being less dense, is forced to rise. This results in the unstable conditions that are characteristic of fronts and that produce much of the precipitation of the continental United States.

FRONTS AND WEATHER MAPS. On weather maps half-circles and/ or triangles are used on the lines representing fronts to indicate the type of front. See Figure 7-8 for specific front symbols. The triangles and half-circles point in the direction the fronts and associated air masses are moving.

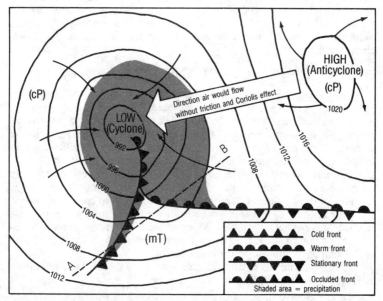

Figure 7-8. Circulation of winds and interaction of high and low pressure air masses in the Northern Hemisphere. Air circulates clockwise in a HIGH and counterclockwise in a LOW.

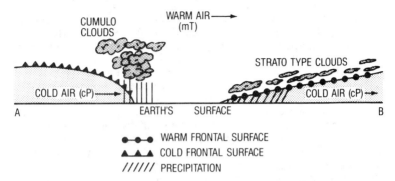

●—●—● WARM FRONTAL SURFACE
▲—▲—▲ COLD FRONTAL SURFACE
////// PRECIPITATION

Figure 7-9. Cross-sectional view of the ---- line in Figure 7-8. The conditions of the cold front at left are: steep slope, cumulo (puffy) clouds, and precipitation just before and after the point where the front meets the earth's surface. The conditions of the warm front at right are: gentle frontal surface, strato (layered) clouds, and a broad band of precipitation preceding the point where the front meets the earth's surface.

TRACKS OF AIR MASSES AND FRONTS. In the United States, the **tracks** (paths) and rate of movement of air masses and fronts can be predicted on the basis of past observations. Thus, the succession of weather changes that accompanies such movements can be reasonably forecasted. The direction of some of these tracks is indicated by the arrows in Figure 7-7. Most of the tracks follow a westerly to easterly route in the United States because of our location in the southwest planetary winds.

WEATHER PREDICTION AND PROBABILITY

Most weather predictions, or forecasts, are based on the **probability** (chance) of occurrence of weather variables based on relationships between weather variables. As an example, suppose that in a twenty-day period the air pressure drops on ten days and that the air temperature increases on eight of those ten days. What is the probability that air temperature will increase if air pressure is decreasing? Based on the data above, there is an 8/10 or 80% chance of increasing temperature. In actual cases of computing probability, the Weather Service uses large amounts of data collected over many years on which to base predictions.

Probabilities have been worked out for a vast number of possible relationships between variables. Many of the relationships are very complex, but some of them are simple direct or inverse relationships. Some of the simple relationships are listed below:

1. Air pressure is related to temperature changes: As temperature increases, pressure decreases, and vice versa.

2. The chance of precipitation increases as the air temperature gets closer to the dew point temperature. This is the case because the smaller the difference between the air temperature and the dew point temperature, the nearer the air is to saturation, condensation, and precipitation.

3. The greater the pressure gradient in an area, the faster is the speed of the winds.

STATION MODELS.   On weather maps the weather conditions for each weather station are shown by symbols arranged in order around a small circle. These symbols and the circle make up a **station model,** which gives the latest readings for all the weather variables. For a sample of a station model see Figure 7-10 below.

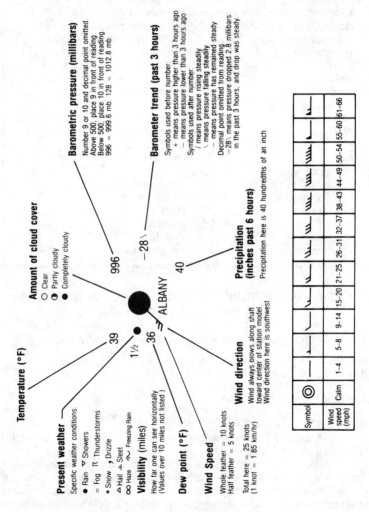

Figure 7-10. Sample Station Model For Weather Map. (Also see the charts of weather map symbols in the *Earth Science Reference Tables*.)

VOCABULARY

weather	relative humidity
atmospheric variables	moisture capacity
isotherm	dew point temperature
condensation	cloud
sublimation	condensation surface
frictional drag	precipitation
adiabatic temperature change	air mass
atmospheric pressure	continental polar air mass (cP)
barometric pressure	continental tropical air mass (cT)
air pressure	maritime tropical air mass (mT)
isobar	maritime polar air mass (mP)
moisture	source region
wind	LOW
pressure gradient	cyclone
convergence	HIGH
divergence	anticyclone
planetary wind belts	front
water vapor	cold front
evaporation	warm front
transpiration	occluded front
evapotranspiration	stationary front
saturation	track
vapor pressure	probability
saturation vapor pressure	visibility
absolute humidity	present weather

QUESTIONS ON TOPIC VII—ENERGY EXCHANGES IN THE ATMOSPHERE AND WEATHER

Questions in Recent Regents Exams (end of book)

Questions from Earlier Regents Exams

1. A balloon carrying weather instruments is released at the earth's surface and rises through the troposphere. As the balloon rises, what will the instruments generally indicate? [Refer to the *Earth Science Reference Tables*.]   (1) a decrease in both air temperature and air pressure   (2) an increase in both air temperature and air pressure   (3) an increase in air temperature and a decrease in air pressure   (4) a decrease in air temperature and an increase in air pressure

2. As a parcel of air rises, its temperature will   (1) decrease due to expansion   (2) decrease due to compression   (3) increase due to expansion   (4) increase due to compression

3. The diagram below shows the direction of movement of air over a mountain.

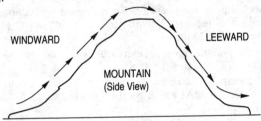

WINDWARD   LEEWARD

MOUNTAIN
(Side View)

As the air moves down the leeward side of the mountain, the air will (1) warm due to compression   (2) warm due to expansion   (3) cool due to compression   (4) cool due to expansion

4. On a clear, dry day an air mass has a temperature of 20°C and a dewpoint temperature of 10°C. According to the graph, about how high must this air mass rise before a cloud can form?   (1) 1.6 km   (2) 2.4 km   (3) 3.0 km   (4) 2.8 km

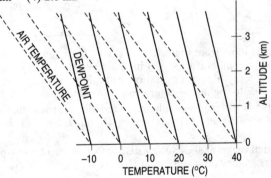

5. According to the graph below, what happens to the density of a mass of air when its water vapor content increases?   (1) density decreases (2) density increases   (3) density remains the same

WATER VAPOR CONTENT

6. At sea level, as the temperature of the atmosphere decreases, the air pressure usually   (1) decreases   (2) increases   (3) remains the same

7. An air pressure of 1023 millibars is equal to how many inches of mercury?   (1) 30.10   (2) 30.15   (3) 30.19   (4) 30.21

8. According to the *Earth Science Reference Tables*, an air pressure of 30.15 inches of mercury is equal to   (1) 1017 mb   (2) 1019 mb   (3) 1021 mb   (4) 1023 mb

9. As the pressure gradient increases, wind velocity   (1) decreases   (2) increases   (3) remains the same

10. The map below represents a portion of an air-pressure field at the earth's surface. At which position is wind speed *lowest*?   (1) A   (2) B   (3) C   (4) D

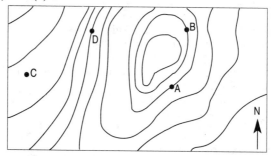

11. Winds blow from regions of   (1) high air temperature to regions of low air temperature   (2) high air pressure to regions of low air pressure   (3) high precipitation to regions of low precipitation   (4) convergence to regions of divergence

12. In the Northern Hemisphere, a wind blowing from the south will be deflected toward the   (1) northwest   (2) northeast   (3) southwest   (4) southeast

13. Which diagram shows the *usual* path followed by low-pressure storm centers as they pass across the United States?

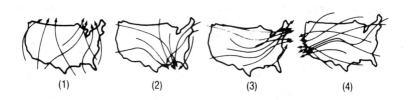

(1)   (2)   (3)   (4)

14. Which is the primary source of energy for ocean currents and waves?   (1) the moon   (2) the atmosphere   (3) the continents   (4) the sun

15. What is the most direct cause of major surface ocean currents?   (1) wind   (2) gravity   (3) tides   (4) pollution

To answer questions 16 through 21 refer to the adjoining diagram, which shows movement of air in the lower part of the atmosphere around the time of an equinox.

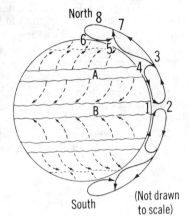

16. The movement of air from 1 to 2 to 3 to 4 to 1 would be called a(n) (1) advection cell   (2) adiabatic cell (3) convection cell   (4) Coriolis cell

17. What is the basic underlying reason for the movement of air shown?   (1) differences in gravity   (2) differences in air density (3) differences in pressure gradient   (4) differences in magnetism

18. How does the movement of air at position 2 compare to the movement of air at position 4?   (1) They both have converging air. (2) They both have diverging air.   (3) Position 2 is diverging and position 4 is converging.   (4) Position 2 is converging and position 4 is diverging.

19. What would be the most likely cause of the movement of air at position 1?   (1) low moisture content and low temperature   (2) high moisture content and high temperature   (3) low moisture content and high temperature   (4) high moisture content and low temperature

20. The air moves at the surface of the earth, from position A to position B, because   (1) positions A and B have low pressure (2) positions A and B have high pressure   (3) position A has low pressure and position B has high pressure   (4) position A has high pressure and position B has low pressure

21. Condensation would most likely occur at position   (1) 3   (2) 4 (3) 6   (4) 7

22. By which process does moisture leave green plants?   (1) convection   (2) condensation   (3) transpiration   (4) radiation

23. Most moisture enters the atmosphere by the processes of (1) convection and conduction   (2) condensation and radiation (3) reflection and absorption   (4) transpiration and evaporation

24. The primary source of most of the moisture for the earth's atmosphere is   (1) soil-moisture storage   (2) rivers and lakes (3) melting glaciers   (4) oceans

25. As the exposed area of a moist object decreases, the rate of evaporation of the liquid from that object   (1) decreases   (2) increases (3) remains the same

26. As the amount of light energy striking a moist object increases, the rate of evaporation of the liquid from that object   (1) decreases   (2) increases   (3) remains the same

27. As the amount of water vapor in a given volume of air increases, the rate of evaporation from a moist object   (1) decreases   (2) increases   (3) remains the same

28. As the rate of molecular activity of a liquid increases, the rate of evaporation of that liquid   (1) decreases   (2) increases   (3) remains the same

29. In the diagram below, at which location would the vapor pressure of the air most likely be greatest?   (1) A   (2) B   (3) C   (4) D

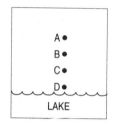

30. Which graph best represents the relationship between the moisture-holding capacity (ability to hold moisture) of the atmosphere and atmospheric temperature?

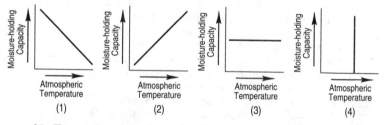

31. To say that the relative humidity of a given day is 70 percent means that the air   (1) is composed of 70 percent water vapor   (2) holds 70 percent of its water vapor capacity   (3) contains 70 parts of water to 100 parts of dry air   (4) contains the same amount of water that it would contain at 70°F

32. During which part of the day is the relative humidity usually *lowest*?   (1) morning   (2) midafternoon   (3) evening   (4) late night

33. Air at a temperature of 60°F and a relative humidity of 51% was warmed to a temperature of 70°F, but the relative humidity remained at 51%. What change occurred in the moisture content?   (1) the moisture content decreased   (2) the moisture content increased   (3) the moisture content remained the same

34. If the amount of water vapor in the air increases, then the dewpoint temperature of the air will   (1) decrease   (2) increase   (3) remain the same

35. The dry-bulb temperature is 20°C. The wet-bulb temperature is 17°C. What is the dewpoint?   (1) 12°C   (2) 13°C   (3) 14°C   (4) 15°C

36. Which statement best explains why the wet-bulb thermometer of a sling psychrometer usually shows a lower temperature than the dry-bulb thermometer?   (1) Water evaporates from the wet-bulb thermometer. (2) Water vapor condenses on the wet-bulb thermometer.   (3) The air around the wet-bulb prevents absorption of heat.   (4) The air around the dry bulb prevents absorption of heat.

To answer questions 37 through 41, refer to the graph below, which shows the hourly surface air temperature, dew point, and relative humidity for a 24-hour period during the month of May at Washington, DC.

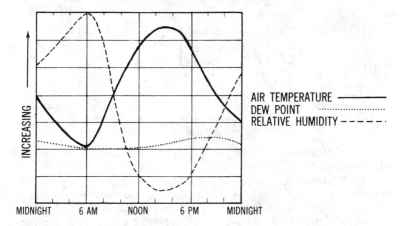

37. The greatest net change in air temperature occurred during the period from   (1) midnight to 6 A.M.   (2) 6 A.M. to noon   (3) noon to 6 P.M.   (4) 6 P.M. to midnight

38. Which conclusion concerning the relationship between the dew point and air temperature is best justified by the data shown? (1) Changes in dew point do not necessarily occur when air temperature changes.   (2) The dew point increases in proportion to an increase in air temperature.   (3) The dew point increases in proportion to a decrease in air temperature.   (4) Changes in dew point are caused by air temperature changes but not in proportion to these changes.

39. The graph indicates that as the air temperature increases, the relative humidity   (1) increases   (2) decreases   (3) sometimes increases and sometimes decreases   (4) remains the same

40. Condensation is most likely to occur at approximately   (1) 6 A.M. (2) 9 A.M.   (3) 7 P.M.   (4) 10 P.M.

41. The energy supply of a hurricane comes from   (1) heat released by condensation   (2) evaporation of tropical waters   (3) the trade winds   (4) discharge of lightning

42. Why do clouds usually form at the leading edge of a cold air mass?   (1) Cold air contains more water vapor than warm air does. (2) Cold air contains more dust particles than warm air does.   (3) Cold air flows over warm air, causing the warm air to descend and cool. (4) Cool air flows under warm air, causing the warm air to rise and cool.

43. Which process most directly results in cloud formation? (1) condensation   (2) transpiration   (3) precipitation   (4) radiation

44. At which temperature could water vapor in the atmosphere change directly into solid ice crystals?   (1) 20°F   (2) 40°F   (3) 10°C (4) 100°C

45. Which graph best represents the relationship between water droplet size and the chance of precipitation?

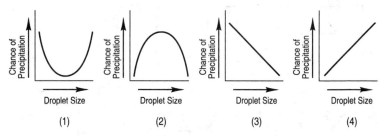

(1)   (2)   (3)   (4)

46. Which process is most likely to remove pollutants from the air? (1) precipitation   (2) evaporation   (3) transpiration   (4) runoff

47. Which statement best explains how atmospheric dust particles influence the water cycle?   (1) Dust particles are the main source of dissolved salts in the sea.   (2) Dust particles increase the capacity of the atmosphere to hold water vapor.   (3) Dust particles increase the amount of evaporation that takes place.   (4) Dust particles provide surfaces on which water vapor can condense.

48. Which graph best shows the relationship between atmospheric transparency and the concentration of pollution particles in the air?

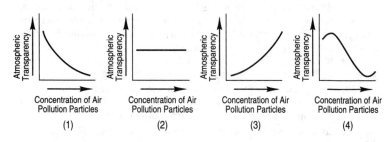

(1)   (2)   (3)   (4)

49. The original characteristics of an air mass are determined by the   (1) surface over which it is formed   (2) pressure of the air mass   (3) insolation it receives   (4) rotation of the earth

50. In order for a large mass of air to acquire uniform characteristics, it must   (1) stagnate over a large land or water surface   (2) descend from the upper troposphere   (3) move rapidly with the prevailing westerlies   (4) move in the general planetary circulation

51. During the summer a warm, moist air mass moved over Texas. This air mass probably originated over   (1) northern Canada   (2) the Pacific Ocean   (3) southern Arizona   (4) the Gulf of Mexico

52. Which best describes the movement of air in a high pressure air mass (anticyclone) in the Northern Hemisphere?   (1) clockwise and away from the center   (2) clockwise and toward the center   (3) counterclockwise and away from the center   (4) counterclockwise and toward the center

53. In approaching the center of a cyclone on a weather map, the numerical values of the isobars   (1) increase   (2) may increase or decrease   (3) decrease   (4) remain the same

54. A high-pressure center is generally characterized by   (1) cool, wet weather   (2) cool, dry weather   (3) warm, wet weather   (4) warm, dry weather

55. The diagram below shows four points on a map with their positions relative to a low-pressure weather system. Which point is most likely having heavy precipitation?

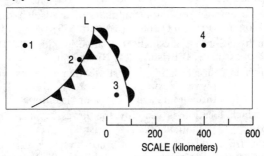

(1) 1   (2) 2   (3) 3   (4) 4

56. The station model below represents atmospheric conditions at a location in New York at noon on a day in January. What is the air pressure indicated on the station model?   (1) 900.3 mb   (2) 1000.3 mb   (3) 1003.0 mb   (4) 9000.3 mb

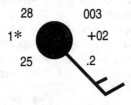

Refer to the weather map below to answer questions 57 through 67.

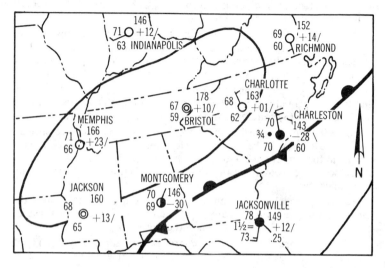

57. Which of the following locations has the highest wind velocity?
(1) Jacksonville   (2) Charlotte   ((3) Indianapolis   (4) Jackson

58. What type of front is found between Montgomery and Jacksonville?   (1) cold   (2) warm   (3) stationary   (4) occluded

59. The barometric reading in millibars of the isobar that passes through Jackson, Miss., is   (1) 160   (2) 1016   (3) 1060   (4) 1600

60. Which station has the highest relative humidity?   (1) Montgomery   (2) Jackson   (3) Bristol   (4) Richmond

61. Which is characteristic of most weather stations within the closed isobar?   (1) low velocity wind   (2) overcast skies   (3) a barometer now lower than three hours ago   (4) poor visibility

62. At which location is visibility the lowest?   (1) Charleston
(2) Jacksonville   (3) Montgomery   (4) Bristol

63. If the front started moving in a northwesterly direction, in a few hours temperatures at Montgomery would most likely   (1) increase   (2) decrease   (3) remain the same

64. At Charlotte, wind is blowing from what compass direction?
(1) South   (2) Northeast   (3) Northwest   (4) Southeast

65. The most precipitation in the six hours before the time of the map was reported at which station?   (1) Memphis   (2) Charleston
(3) Jacksonville   (4) Richmond

66. The air mass found south and east of the front would most likely be classified as a(n)   (1) cP   (2) cT   (3) mT   (4) mP

67. What was the barometric pressure at Indianapolis 3 hours before the time of the map?   (1) 1014.6 mb   (2) 1015.8 mb
(3) 1013.4 mb   (4) 1026.6 mb

Additional Questions

1. Which is *not* true about wind?   (1) Wind direction is named for the direction toward which the wind blows.   (2) Wind moves from regions of higher pressure to lower pressure.   (3) The steeper the gradient, the greater the wind speed.   (4) Wind is horizontal movement of air.

2. When the dry-bulb temperature is 20°C and the wet-bulb temperature is 16°C, the percent relative humidity is   (1) 42%   (2) 62%   (3) 66%   (4) 69%

3. When the dry-bulb temperature is 24°C and the difference between the wet-bulb and dry-bulb temperatures is 10°C, the percent relative humidity is   (1) 8%   (2) 9%   (3) 30%   (4) 54%

4. If at sea level the temperature is 20°C and the dewpoint is 16°C, what is the most likely altitude of the bottom of the local clouds? (1) 0.5 km   (2) 1 km   (3) 1.8 km   (4) 3 km

5. If at an altitude of 1 km the temperature is 10°C, what would be the temperature of the troposphere at 2 km assuming that no condensation of water occurred?   (1) – 10°C   (2) 0°C   (3) 10°C   (4) 20°C

To answer questions 6–13 use the diagram below, which shows four station models, at the same instant in time, identified by letters *A–D*. The stations are located on a west-end line at fifty-mile intervals in the United States.

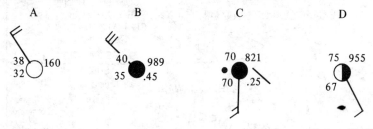

6. The strongest wind is reported at station   (1) *A*   (2) *B*   (3) *C* (4) *D*

7. Station *B* has just experienced the passage of a   (1) warm front (2) cold front   (3) stationary front   (4) hurricane

8. At station *D* the wind is blowing from the   (1) northeast (2) northwest   (3) southeast   (4) southwest

9. The temperature at station *A* is   (1) 6°F   (2) 16°F   (3) 32°F (4) 38°F

10. The relative humidity at station *C* is   (1) 0%   (2) 25%   (3) 70% (4) 100%

11. Heaviest rainfall in the past 6 hours was reported from station (1) *A*   (2) *B*   (3) *C*   (4) *D*

12. Clear skies are reported from station   (1) *A*   (2) *B*   (3) *C* (4) *D*

13. What is the wind speed at station *B* in km/h?   (1) 3 km/h (2) 5.5 km/h   (3) 30 km/h   (4) 55.5 km/h

TOPIC VIII — Water, Energy, and Climate

THE WATER CYCLE

The **water cycle** is a model used to illustrate the movement and the phase changes of water at and near the earth's surface (see Figure 8-1). Below are some of the important aspects of the water cycle:

1. The ultimate source of most water on land is the oceans.

2. The moisture gets to the land from the oceans by way of the atmosphere.

3. When precipitation falls on the land, four things can happen to it:

a. Storage. It can be stored on the land surface as ice or snow.

b. Infiltration. It can infiltrate, or seep into, the earth. The water beneath the earth's surface is called **subsurface water.**

c. Runoff. It can flow on the surface of the land, and is then called **runoff.**

d. Evapotranspiration. Usually a large percentage of precipitation is evaporated or transpired back into the atmosphere.

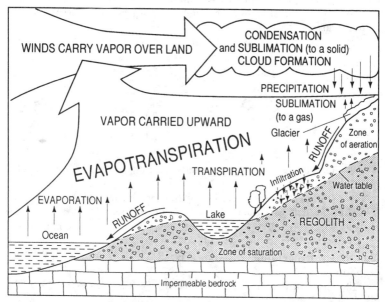

Figure 8-1. The water cycle: a model of the movement and phase changes of water at and near the earth's surface. The water table is an interface of changing position between the zone of saturation, where the pores of the regolith are filled with water (ground water), and the zone of aeration, where the pores are only partly filled with water (capillary water).

FACTORS AFFECTING INFILTRATION. Most infiltration occurs in the *regolith*, which is the unconsolidated material (including soil) at the earth's surface. The amount of water that can infiltrate when precipitation occurs depends on several variables:

1. *The slope of the land.* The steeper the slope (gradient), the less the infiltration.

2. *The degree of saturation of the regolith.* The more saturated the regolith, the less the infiltration. Figure 8-1 shows that the regolith is divided into two zones: the **zone of saturation,** where the pores between solid particles are filled with water; and the **zone of aeration,** where the air spaces are partly filled with water (capillary water). Water infiltrates until it meets the interface between the zone of saturation and the zone of aeration. This interface is called the **water table.** The height of the water table varies with the amount of infiltration. The portion of the subsurface water below the water table is called **ground water.**

3. *The porosity of the regolith.* **Porosity** is the percentage of open space (pores) in a material compared to its total volume. Generally, the greater the porosity of the regolith, the greater the amount of infiltration that can occur.

The porosity of a material is determined by the *shape, packing,* and *sorting* of the particles composing it. (See Figure 8-2.)

a. *Shape.* Well-rounded particles have greater porosity than particles with angular shapes, because the round ones do not fit together so well.

b. *Packing.* The more closely packed the particles, the lower the porosity.

c. *Sorting.* If all the particles in a material are about the same size, they are called **sorted;** if the particles are of mixed sizes, they are **unsorted.** The more unsorted the particles, the lower the porosity, because the small particles can fit into the spaces between the larger particles. It should be noted that size by itself does not affect porosity. For example, a material with large particles may have about the same porosity as one with smaller particles if the shape, packing, and sorting of both particle sizes are about equal.

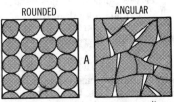

Well-rounded particles have more porosity than particles with angular shapes.

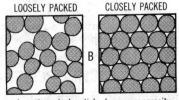

Loosely packed particles have more porosity than closely packed particles.

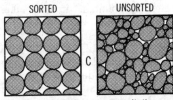

Sorted particles have more porosity than unsorted particles.

Figure 8-2. Effect of particle or sediment shape, packing, and sorting on porosity of the regolith.

4. *The permeability of the regolith.* **Permeability** is the ability of a material to allow fluids such as water to pass through it. The **permeability rate** is the speed at which fluids will flow through a material. A material can be porous and yet impermeable (not permeable). Impermeability may be due to tight packing or cementing of particles, which seals off the pores from one another so that water cannot enter them. In the winter the cementing is often due to ice. In loose particles, the larger the particle size, the faster the permeability rate. The reason is that the size of the pores increases with the size of the particles, thus reducing the amount of friction.

5. *Capillarity.* During infiltration, some water is stopped from moving downward by the attractive force between water molecules and the surrounding earth materials. This attractive force is called **capillarity.** The water that is thus stored in small openings in the zone of aeration is called **capillary water.**

Capillarity also causes water to move up from the water table toward the earth's surface in the zone of aeration. This upward movement is called **capillary migration (capillary action).**

When the particle size of loose particles becomes smaller, the capillarity becomes greater.

FACTORS AFFECTING RUNOFF AND STREAM DISCHARGE. Surface runoff can occur when (1) rate of precipitation exceeds the permeability rate (or infiltration rate) of the earth's surface, (2) the regolith pore space is saturated, or (3) the slope of the surface is too great to allow infiltration to occur.

Most runoff gets to streams, which eventually carry the water to the oceans. The greater the runoff, the greater the amount of *stream discharge* in local streams. **Stream discharge** is the volume of water flowing past a certain spot in a stream in a specific amount of time, and is expressed in such units as cubic meters/second.

WATER, OR HYDROSPHERIC, POLLUTION

SOURCES OF WATER POLLUTANTS. Pollutants are added to the hydrosphere by individuals, community action, and industrial processes. Generally, as the population density of an area increases, the concentration of pollutants in local lakes, streams (rivers), and ground water increases.

TYPES OF POLLUTANTS IN THE HYDROSPHERE. Water pollutants include dissolved and suspended materials, organic and inorganic wastes, outflow from industrial processes, radioactive substances, larger than normal concentrations of certain life forms, and thermal (heat) energy.

Excessive heat energy is considered a source of pollution because of its effects on oxygen content of water. The higher the temperature of water, the less the amount of dissolved oxygen it can hold. Furthermore, high temperatures increase the activity of the **aerobic bacteria** that are normally present. These bacteria use oxygen for their life processes, and their increased activity reduces the oxygen content of the water still further, thus killing many of the other forms of life that are present. Eventually, the aerobic bacteria are replaced by **anaerobic** varieties (types of bacteria that live without free oxygen). These bacteria and their wastes then become pollutants.

Power generating plants, including nuclear reactor types, are a major source of *thermal* (heat) pollution of bodies of water.

LONG-RANGE EFFECTS OF WATER POLLUTION. As water pollution increases, more and more water becomes unfit for human use. Purification of unfit water is a very complex and costly process.

LOCAL WATER BUDGET

The **local water budget** is a numerical model of a local area's water supply by months during a year. The water budget is useful to help determine an area's climate type and has practical applications in flood and irrigation control.

SYMBOLS AND MEANING OF TERMS IN LOCAL WATER BUDGET. To understand the local water budget, one must know the meaning of the terms used and how they relate to one another. The list below shows the symbols and the meaning of each of the terms used in the local water budget. Quantities of water per unit of area are expressed in millimeters. This is the depth to which the water would fill a pan covering the area involved.

1. Precipitation *(P)*. The moisture source for the local water budget is precipitation.

2. Potential evapotranspiration (E_p) is the amount of water that would evapotranspirate if the water were available. The E_p of an area is directly related *(a)* to the amount of energy (heat) available to the area (which for most places is largely determined by the amount of insolation—thus E_p generally decreases as latitude increases), and *(b)* the amount of evaporation surface area. An area of forest has more evaporation surface area for evapotranspiration, thus it has a higher E_p than an equal area of lawn receiving the same insolation. Figures 8-3 and 8-4 on page 110 show the relationships between insolation and latitude and between insolation and potential evapotranspiration.

3. The difference between precipitation and potential evapotranspiration $(P - E_p)$. Its value can be negative, positive, or zero.

4. The amount of change in the water stored in the ground (ΔSt). Its value can be negative, positive, or zero. If ΔSt is positive, it is called **recharge**. Recharge occurs when the soil is not saturated and P is greater than E_p. Negative ΔSt is called **usage**. Water usage occurs when E_p is greater than P and soil storage is greater than zero.

5. Soil storage *(St)* is the amount of water stored in the soil. Each area has a specific maximum amount of soil storage, but for most areas it is assumed to be 100 millimeters.

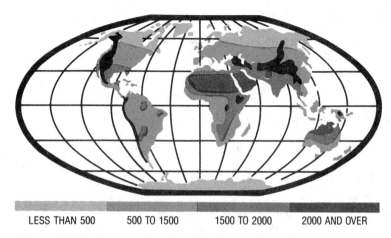

LESS THAN 500   500 TO 1500   1500 TO 2000   2000 AND OVER

Figure 8-3. World map of insolation at the earth's surface in thousands of calories/cm²/year. Note how yearly insolation generally increases with decreasing latitude. Due to many factors, however, especially differences in cloud coverage, it is not a perfect correlation. (Darkest areas have very high elevation.)

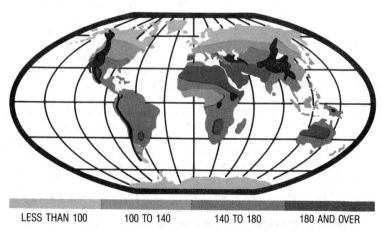

LESS THAN 100   100 TO 140   140 TO 180   180 AND OVER

Figure 8-4. World map of potential evapotranspiration in centimeters per year. Note the very good comparison to the yearly insolation map above. It is not a perfect correlation, because local heat values are affected by many other factors such as surface ocean currents and tropospheric convection currents. (Darkest areas have very high elevation.)

6. Actual evapotranspiration (E_a). E_a can never be greater than E_p. When there is a deficit, E_a is less than E_p, and at all other times E_a equals E_p.

7. Water deficit (D). A water deficit exists when E_a does not equal E_p because there is not enough water in the combination of P and St. The deficit is the amount of water that would be necessary for E_a to equal E_p, that is, it is E_p minus E_a. A deficit can occur only when the combined St and P are not as great as E_p.

8. Water surplus (S). There is a surplus when the soil is saturated (maximum St) and P is greater than E_p. Surplus water eventually becomes runoff.

EXAMPLES OF LOCAL WATER BUDGETS. To understand how a water budget is worked out, study the procedure that is described for the water budget of Syracuse, New York (Figure 8-6 on page 112). Then use the procedure to complete the water budget for El Paso, Texas (Figure 8-7 on page 113). Graph your results and see if they agree with the graph in Figure 8-7.

STREAM DISCHARGE AND WATER BUDGET. The discharge of a local stream is usually a measure of the amount of surplus water in the local drainage area of the stream. Figure 8-5 shows the discharge of a stream in Syracuse, New York. Comparison of this data with the water budget of Syracuse shows that the stream discharge is greatest when the surplus is the largest and least when the surplus is smallest.

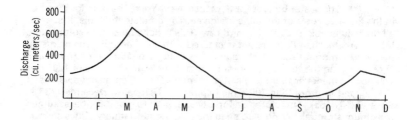

Figure 8-5. Discharge of a stream in Syracuse, New York, which is dependent on local precipitation. Compare with the water budget of Syracuse (Fig. 8-6). Much of the surplus in the winter months is stored temporarily as snow and ice, and does not contribute to stream discharge until early spring. The discharge during the period of deficit (August and September) is due to depletion of ground water at levels too deep to be available to plants.

Note that the stream is still flowing when Syracuse has a deficit in the water budget. If a local stream is still flowing during a deficit period, the source of the water is ground water depletion.

	J	F	M	A	M	J	J	A	S	O	N	D	Totals
P	72	68	81	75	74	87	84	82	72	76	68	72	911
E_p	0	0	3	34	83	115	134	122	84	46	15	0	636
$P-E_p$	72	68	78	41	−9	−28	−50	−40	−12	30	53	72	
ΔSt	0	0	0	0	−9	−28	−50	−13	0	30	53	17	
St	100	100	100	100	91	63	13	0	0	30	83	100	
E_a	0	0	3	34	83	115	134	95	72	46	15	0	597
D	0	0	0	0	0	0	0	27	12	0	0	0	39
S	72	68	78	41	0	0	0	0	0	0	0	55	314

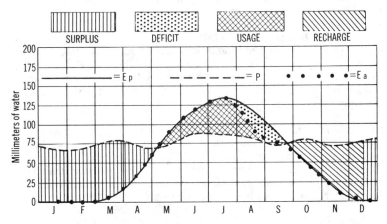

Figure 8-6. The water budget for Syracuse, New York. The data for P and E_p in the first two rows of the table are given. To complete the table, first enter all the values of P − E_p in the third row. This is done by simple subtraction. If E_p is greater than P, the result is entered with a minus sign. The next step is to find a month in which storage (St) is either 0 or 100. When the total P is greater than the total E_p (a humid climate), look for a storage of 100. This will occur whenever there is a series of months in which positive values of P − E_p add to more than 100. This happens in October to December (30 + 53 + 72 = 155). Therefore, St for December must be 100 (the assumed maximum storage). We now go on to the next month, January. Since storage is full, and P − E_p is positive, there is no change in storage (ΔSt = 0), and all of P − E_p (72 mm) is surplus. We also make E_a equal to E_p. (It is zero in this month because the ground is frozen and plants are dormant.) For the next three months, St remains at 100, E_a = E_p, and all of P − E_p is surplus.

In May, P − E_p is negative. However, the difference can come out of storage. Thus, ΔSt = −9, St becomes 91, and E_a = E_p because there is enough water available. But there is no surplus this month. This continues until August. In August, P − E_p = −40, but there is only 13 mm in storage. Therefore E_a is only 95 (P of 82 plus 13 from St). St drops to zero, and there is a deficit of 27 (E_p − E_a). In September there is another deficit. E_a equals P, because there is no water in storage. In October, P − E_p becomes positive, and recharge begins (ΔSt = 30). In November, ΔSt = 53, bringing St up to 83. In December, St becomes 100 (as already noted), and there is a surplus of 55.

	J	F	M	A	M	J	J	A	S	O	N	D	Totals
P	10	11	8	7	8	18	40	41	33	17	13	12	218
E_p	11	18	37	75	117	163	171	152	111	62	24	10	951
$P-E_p$	-1	-7	-29	-68	-109	-145	-131	-111	-78	-45	-11	2	
ΔSt													
St				0									
E_a													
D													
S													

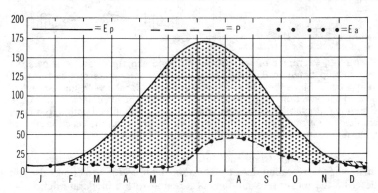

Figure 8-7. The water budget for El Paso, Texas. Since P is less than E_p, this is an arid climate. To complete the table, look for a series of months in which negative values of $P - E_p$ add to -100. From January to April, the total is -105. Therefore, St for April must be 0. This can be your starting point.

CLIMATE

Climate is the overall view of weather conditions of a region over a long period of time. The two major aspects of climate are *temperature* and *moisture conditions*. The factors that determine climate include latitude, planetary wind and pressure belts, oceans and other large bodies of water, ocean currents, mountains, and elevation.

TEMPERATURE AND CLIMATE.   In terms of climate, two characteristics of the temperatures of a region are important: (1) the average temperature over the year, and (2) the range of average monthly temperatures from lowest to highest during the year.

MOISTURE AND CLIMATE.   A climate is called **arid,** or dry, if the total precipitation during the year is less than the potential evapotranspiration, that is, if P is less than E_p. A climate is called **humid** if P is greater than E_p. Thus, whether a climate is said to be arid or humid depends not on the amount of precipitation but on the difference between the amount of moisture available and the potential need. As illustrated in Figure 8-8 on page 114, a region can have very little precipitation and still have a humid climate.

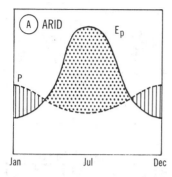
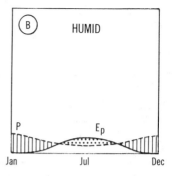

Figure 8-8. Examples of an arid and a humid climate. Graph A shows the relationship between P and E_p for a region of moderate rainfall but high values of E_p. Since there is a net moisture deficit for the year, the region has an arid climate. Some regions around 30° latitude have climates of this type. Graph B represents a climate in which there is little precipitation, but even smaller values of E_p. Since there is a moisture surplus for the year, the climate is humid. Many polar regions have climates of this type.

Regions with extreme moisture deficits (large yearly negative values of $P - E_p$) are called **deserts.** The term *semi-arid* is sometimes used to describe the moisture conditions of a region with a moderate deficit.

LATITUDE AND TEMPERATURE.   Latitude is a major factor in determining climates because of its influence on both temperature and moisture conditions. Temperature characteristics vary with latitude because of the relationships between insolation and latitude discussed in Topic VI. At low latitudes, where the maximum angle of insolation is always high, average temperatures are high throughout the year. Because the duration of insolation is fairly constant at about 12 hours per day, there is little temperature variation during the year. At high latitudes, where the maximum angle of insolation is never large and in some months remains zero, average temperatures are low. Since the duration of insolation varies from zero to 24 hours a day, and is longest at the times of greatest angle of insolation, temperatures vary over a wide range from winter lows to summer highs.

LATITUDE AND MOISTURE.   Moisture conditions vary with latitude because of the location of the planetary wind and pressure belts (see page 85). Where there is low pressure, such as near the equator and in mid-latitudes, the air rises. The cooling of that rising air results in large amounts of precipitation and humid climates. When air falls in high-pressure belts, such as around 30° latitude, it warms and there is little precipitation. Such areas have arid climates.

LATITUDINAL CLIMATE PATTERNS.   The combination of the temperature and moisture effects of latitude results in a basic distribution of climate types around the world, called **latitudinal climate patterns** (see Figure 8-9).

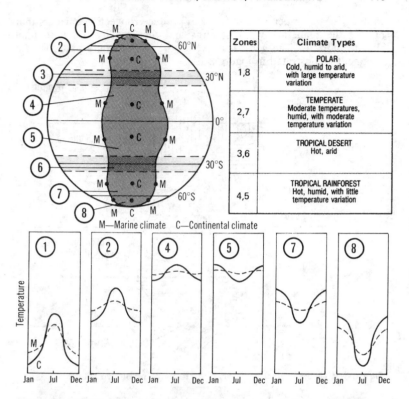

Figure 8-9. Basic latitudinal climate pattern on an imaginary continent. The arid belts around 30° latitude are the result of descending air in the atmospheric circulation at that latitude (see Figure 7-3). The modification of this basic pattern by prevailing winds and distance from oceans is illustrated in Figure 8-10.

EFFECT OF LARGE BODIES OF WATER ON CLIMATE. Large bodies of water serve to modify the latitudinal climate patterns. If an area is near the ocean or a large lake, its temperatures will be moderated by the slow heating up and cooling off of the water body. An area with these moderated temperatures is said to have a **marine climate**. Marine climates have cooler summers and warmer winters, thus a small annual range of temperatures compared to inland areas at the same latitude. Inland areas away from large bodies of water have cooler winters and warmer summers and a large annual range of temperatures. Such areas are said to have a **continental climate**.

EFFECT OF LARGE LANDMASSES ON CLIMATE. As explained on page 67, bodies of land heat up and cool off more rapidly than bodies of water. As a result, the temperatures over a continent in the mid-latitudes are higher than the adjacent oceans in summer and lower in winter. These temperature differences produce convection currents that modify the planetary wind patterns and thus modify the latitudinal climate patterns. The general effect on the basic climate patterns of a continent is illustrated in Figure 8-10.

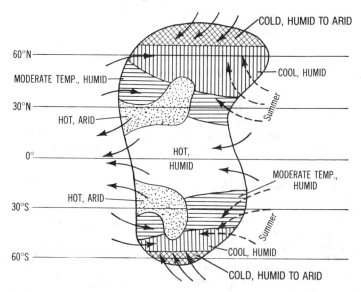

Figure 8-10. Modification of basic climate patterns by prevailing winds.

Notice in Figure 8-10 that the arid belt around 30° latitude does not extend across the eastern portion of the continent. The reason for this is that in the summer months the land temperatures are higher than the ocean temperatures, causing air to rise over the land and resulting in prevailing winds carrying moisture inland from the oceans on the east. The eastern regions of the continent therefore have much more precipitation than they would have if the planetary wind pattern remained unchanged.

A second effect is the arid region in the central portion of the continent at the mid-latitudes (40° to 50°). The reason for this is that the prevailing winds from the western oceans lose their moisture by precipitation and become increasingly dry as they blow inland.

STORM TRACKS AND CLIMATE.   In the mid-latitudes, such as the continental United States, temperature and moisture are greatly affected by a succession of low-pressure systems. Major tracks of low-pressure storm systems are shown in Figure 8-11.

OCEAN CURRENTS AND CLIMATE.   Coastal climate patterns are modified by ocean currents. Currents flowing away from the equator carry warm water to higher latitudes, while currents flowing toward the

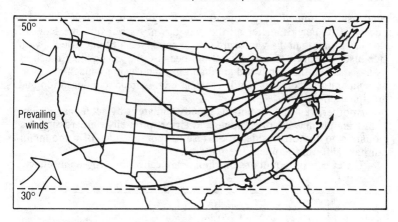

Figure 8-11. The usual paths followed by low-pressure storm centers as they pass across the United States.

equator carry cool water to lower latitudes. A cool ocean current will cause a coastal area to have cooler temperatures and less precipitation. One reason there is low precipitation is that cool water results in cool air, which has a low water vapor capacity and thus cannot hold much water vapor. The locations of the warm and cool ocean currents are shown on the *Surface Ocean Currents* map in the *Earth Science Reference Tables.*

ELEVATION AND CLIMATE.   The elevation of an area above sea level modifies the latitudinal climate pattern, because as air rises, it expands and cools (at the rate of 10°C/km; see *Lapse Rate* in the *Earth Science Reference Tables*). Thus the higher the altitude at any given latitude, the cooler it is. Elevation also affects precipitation. As the elevation increases, the temperature and water vapor capacity decrease; the air thus approaches the dew point. Therefore, areas at higher altitudes generally have more precipitation than lower areas.

MOUNTAINS AND CLIMATE—OROGRAPHIC EFFECT.   Mountains that intersect prevailing winds, such as those associated with the planetary wind belts, can modify the latitudinal climate pattern in the following manner. Figure 8-12 on page 118 shows a cross section of a mountain against which the prevailing winds blow from the left. This side is called the *windward side*. As the wind strikes the windward side of the mountain, the air is forced to rise. As it does so, it cools adiabatically. As long as the air temperature is above the dew point, the dry adiabatic cooling rate is about 10°C per kilometer. If the cooling continues until the dew point is reached, condensation will occur above this altitude. Since large amounts of latent heat are released during condensation, the temperature of the rising air does not drop as rapidly as before. The adiabatic cooling rate during condensation is only 6°C per kilometer.

On the opposite side of the mountain, called the *leeward side*, the air begins to descend. It therefore begins to warm adiabatically. Since condensation immediately stops, the warming rate is constant at the dry rate of 10°C per kilometer all the way down. As a result, the leeward side is warmer than the windward side at any given altitude. The leeward side also has much less precipitation, because the air has lost much of its moisture on the windward side, and its water vapor capacity rises as its temperature increases. Both of these factors make condensation and precipitation unlikely.

Another way that mountains modify climate is by acting as barriers to moving air masses, preventing cold air or warm air from crossing the mountain to the other side. As a result, opposite sides of a mountain can have different temperature patterns.

The effects of mountains on climate are called the **orographic effect.**

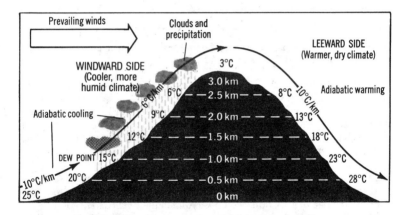

Figure 8-12. Orographic climate effect of mountains that intersect prevailing moist winds (such as those on the northwest coast of the United States). Compare the temperature at the same altitude on the two sides of the mountain. (See text for detailed explanation.)

CLIMATIC CLASSIFICATION BASED ON PRECIPITATION AND POTENTIAL EVAPOTRANSPIRATION. Climatic regions or zones can be distinguished qualitatively based on either P/E_p (precipitation divided by potential evapotranspiration) or $P - E_p$ (precipitation minus potential evapotranspiration). The classification can be based on a system like the one in the table below.

Climate Type	P/E_p	$P - E_p$ in cm
Humid	1.2 or greater	26.7 or greater
Sub-humid	1.2 to 0.8	26.7 to 0.0
Semi-arid	0.8 to 0.4	0.0 to -26.7
Arid	0.4 or lower	-26.7 or lower

VOCABULARY

water cycle	local water budget
infiltration	precipitation (P)
subsurface water	potential evapotranspiration (E_p)
runoff	change in soil storage (ΔSt)
zone of saturation	recharge ($+\Delta St$)
zone of aeration	usage ($-\Delta St$)
water table	soil storage (St)
ground water	actual evapotranspiration (E_a)
porosity	deficit (D)
sorted and unsorted particles	surplus (S)
permeability	climate
permeability rate	arid climate
capillarity	humid climate
capillary water	desert
capillary migration (action)	latitudinal climate pattern
stream discharge	marine climate
aerobic bacteria	continental climate
anaerobic bacteria	orographic effect

QUESTIONS ON TOPIC VIII—WATER, ENERGY, AND CLIMATE

Questions in Recent Regents Exams (end of book)

June 1995:   25, 27, 66, 67, 70, 76–80
June 1996:   15, 23, 26–29, 76–80, 101
June 1997:   23–25, 27, 28, 78
June 1998:   23–27

Questions from Earlier Regents Exams

1. For the earth as a whole, what happens to most of the precipitation?   (1) It recharges the soil moisture deficit.   (2) It becomes runoff and moves to the oceans.   (3) It is stored in the soil as capillary water.   (4) It is returned to the atmosphere through evapotranspiration.

2. The primary source of most of the moisture for the earth's atmosphere is   (1) soil-moisture storage   (2) rivers and lakes   (3) melting glaciers   (4) oceans

3. As the amount of precipitation on land increases, the depth from the surface of the earth to the water table will probably   (1) decrease   (2) increase   (3) remain the same

4. As the heat and pressure applied to buried sediments is increased, the porosity of the sediments will probably   (1) decrease   (2) increase   (3) remain the same

5. Which diagram best illustrates the condition of the soil below the water table?

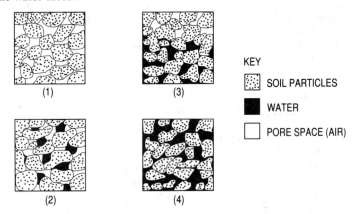

6. Which graph best represents the relationship between porosity and particle size for soil samples of uniform size, shape, and packing?

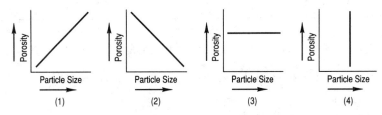

7. Which graph best represents the relationship between soil permeability rate and infiltration when all other conditions are the same?

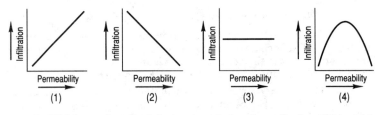

8. Which earth material covering the surface of a landfill would permit the *least* amount of rainwater to infiltrate the surface?   (1) silt   (2) clay   (3) sand   (4) pebbles

9. As the temperature of the soil decreases from 10°C to −5°C, the infiltration rate of ground water through this soil will most likely (1) decrease   (2) increase   (3) remain the same

10. Surface runoff of precipitation occurs when   (1) porosity is exceeded by permeability   (2) the infiltration rate is greater than the precipitation rate   (3) the precipitation rate is greater than the infiltration rate   (4) there is no demand for evapotranspiration

11. Which graph best illustrates the relationship between the slope of the land and the amount of surface runoff?

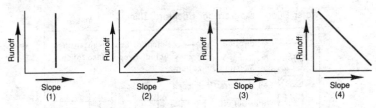

12. Which graph best represents the typical relationship between population density near a lake and pollution of the lake?

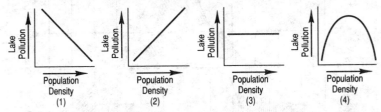

13. What is the main reason that a high concentration of aerobic bacteria is harmful to a lake? (1) The bacteria release large amounts of oxygen. (2) The bacteria use up large amounts of oxygen. (3) The bacteria cause excessive cooling of the water. (4) The bacteria provide food for predators.

14. During a given month in a certain locality, the potential evapotranspiration exceeded the actual evapotranspiration. Which probably occurred? (1) a water deficit (2) a water surplus (3) a higher potential evapotranspiration than usual (4) low temperatures and high humidity

15. Which would cause the potential evapotranspiration to decrease in a given month? (1) below-normal precipitation (2) drilling of a large well (3) a month-long cold spell (4) a high actual evapotranspiration

16. The main source of moisture for the local water budget is (1) potential evapotranspiration (2) actual evapotranspiration (3) ground water storage (4) precipitation

17. The actual evapotranspiration (E_a) during the year is (1) always greater than potential evapotranspiration (E_p) (2) sometimes greater than potential evapotranspiration (E_p) (3) never greater than potential evapotranspiration (E_p)

Base your answers to questions 18 through 24 on the water budget diagram for Rockford, Illinois, below.

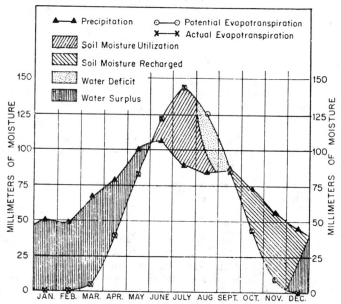

Water Budget Diagram for Rockford, Illinois

18. A water deficit occurred in Rockford, Illinois, during   (1) February   (2) June   (3) August   (4) December

19. During how many months of the year did Rockford have a water surplus?   (1) ten   (2) eight   (3) six   (4) four

20. During which of these months did the potential evapotranspiration equal the precipitation in Rockford?   (1) March   (2) January   (3) September   (4) November

21. During which month was the soil moisture recharge the greatest?   (1) December   (2) November   (3) October   (4) August

22. If the storage capacity of Rockford were increased from 100 to 150 mm, which would not change?   (1) actual evapotranspiration   (2) potential evapotranspiration   (3) deficit   (4) surplus

23. During which month are plants most likely in need of irrigation?   (1) April   (2) June   (3) August   (4) October

24. Streams in this area probably carry the greatest volume of water during the month of   (1) April   (2) June   (3) August   (4) November

25. Dry soil will be recharged with moisture if potential evapotranspiration is   (1) less than precipitation   (2) greater than precipitation   (3) equal to precipitation

26. During a 3-week period without rain in June, water continued to flow in a small New York State stream. The water in the stream most likely came from   (1) the roots of trees along the stream bank   (2) evapotranspiration in a region far away and unaffected by the dry period   (3) ground water flowing into the streambed   (4) condensation on the surface of rocks in the stream

27. A region with an arid climate has an annual precipitation that is (1) less than the annual potential evapotranspiration   (2) greater than the annual potential evapotranspiration   (3) equal to the annual potential evapotranspiration

28. Which generally has the greatest effect in determining the climate of an area?   (1) degrees of longitude   (2) extent of vegetation cover   (3) distance from equator   (4) month of the year

29. As the number of degrees of latitude from the equator increases, the yearly average temperature generally   (1) decreases   (2) increases (3) remains the same

Refer to the diagrams to answer question 30. Diagram I shows the planetary wind belts of the earth. Diagram II is a graph of the average yearly precipitation for locations between 90° north latitude and 90° south latitude.

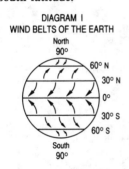

DIAGRAM I
WIND BELTS OF THE EARTH

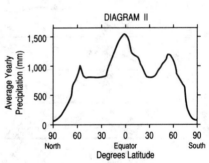

DIAGRAM II

30. Which statement about the earth's average amount of yearly precipitation is best supported by the diagrams?   (1) Precipitation is lowest at latitudes where planetary winds converge (meet).   (2) Precipitation is highest at latitudes where planetary winds converge (meet). (3) Precipitation is highest at latitudes where planetary winds diverge (move apart).   (4) Precipitation is unrelated to planetary wind belts.

31. Although New York City is at approximately the same latitude as Omaha, Nebraska, its winter months are warmer and its summer months cooler. Which statement best explains why this is so?   (1) The sun's rays are more direct on New York City in the winter.   (2) Nebraska is nearer the Rocky Mountains.   (3) The water around New York City has a moderating effect on the temperature.   (4) The prevailing westerlies have a greater effect on Omaha than on New York City.

32. Which area of New York State would probably have the lowest annual temperature range?   (1) Long Island   (2) the Catskills (3) the Adirondack peaks   (4) the Mohawk Valley

33. Bodies of water have a moderating effect on climate primarily because   (1) water gains heat more rapidly than land does   (2) water surfaces are flatter than land surfaces   (3) water temperatures are always lower than land temperatures   (4) water temperatures change more slowly than land temperatures do

34. Which locality has the greatest annual range of temperature? (1) Seattle, Washington   (2) Bismarck, North Dakota   (3) New York City   (4) Miami, Florida

35. A low-pressure storm center located over New York State will most likely move toward the   (1) southeast   (2) southwest   (3) northeast   (4) northwest

36. As a parcel of air rises, its temperature will   (1) decrease due to expansion   (2) decrease due to compression   (3) increase due to expansion   (4) increase due to compression

Base your answers to questions 37–42 on your knowledge of earth science and on the diagram. The diagram represents an imaginary continent on the earth surrounded by water. The arrows indicate the direction of the prevailing winds. Two large mountain regions are also indicated. Points *A*, *B*, *E*, and *H* are located at sea level; *C*, *D*, and *F* are in the foothills of the mountains; *G* is high in the mountains.

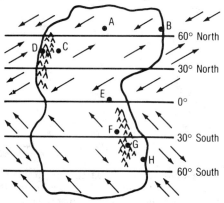

37. Which physical characteristic would cause location *G* to have a colder yearly climate than any other location?   (1) the nearness of location *G* to a large ocean   (2) the location of *G* with respect to the prevailing winds   (3) the elevation of location *G* above sea level (4) the distance of location *G* from the equator

38. Which location probably has the greatest annual rainfall?   (1) *A* (2) *F*   (3) *C*   (4) *D*

39. Which location would be warmed by adiabatic changes?   (1) *A* (2) *B*   (3) *C*   (4) *E*

40. Which graph best represents the average monthly temperatures that would be recorded during one year at location *E*?

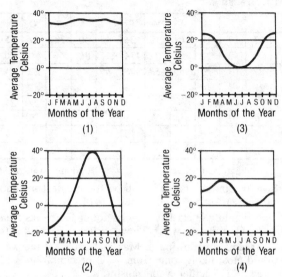

41. Which location probably has the greatest range in temperature during the year?   (1) *A*   (2) *B*   (3) *H*   (4) *D*

42. Which location will probably record its highest potential evapotranspiration values for the year during January?   (1) *A*   (2) *F*   (3) *C* (4) *D*

The diagram below shows the positions of the cities of Seattle and Spokane, Washington. Both cities are located at approximately 48° north latitude, and they are separated by the Cascade Mountains.

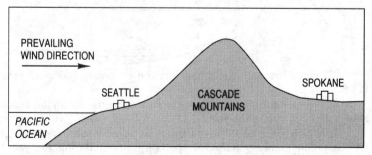

43. How does the climate of Seattle compare with the climate of Spokane?   (1) Seattle—hot and dry; Spokane—cool and humid (2) Seattle—hot and humid; Spokane—cool and dry   (3) Seattle—cool and humid; Spokane—warm and dry   (4) Seattle—cool and dry; Spokane—warm and humid

Additional Questions

Complete the water budget for Fairbanks, Alaska, given below to answer questions 1 through 4.

	J	F	M	A	M	J	J	A	S	O	N	D
P	27	13	10	6	18	36	48	60	28	22	15	13
E_p	0	0	0	0	71	114	115	86	48	0	0	0
$P-E_p$	27	13	10	6	−53	−78	−67			22	15	13
ΔSt	27	13	10							22	15	13
St	77	90	100							22	37	50
E_a	0	0	0							0	0	0
D	0	0	0							0	0	0
S	0	0	0							0	0	0

Values in millimeters of water   Maximum storage: 100 mm

1. Stream flow discharge would be the greatest in the month of (1) February   (2) April   (3) October   (4) December

2. How would you classify the climate of Fairbanks from June to September?   (1) rain forest   (2) low-latitude   (3) arid   (4) moist

3. If streams are flowing in May in Fairbanks, the water in the streams would most likely come from   (1) precipitation   (2) surplus   (3) soil storage   (4) actual evapotranspiration

4. If the porosity of the soil in this area increases, which of the following would change the most?   (1) precipitation   (2) potential evapotranspiration   (3) soil storage   (4) precipitation minus potential evapotranspiration

Refer to the diagram below to answer questions 5 through 8.

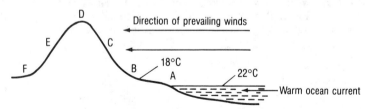

5. Which location would have the coolest average temperature? (1) A   (2) C   (3) D   (4) F

6. Which location would have the most condensation in the atmosphere above it during the year?   (1) B   (2) C   (3) E   (4) F

7. Which place would have the largest negative value for $P - E_p$ (precipitation minus potential evapotranspiration) during the year? (1) A   (2) B   (3) D   (4) F

8. How is the potential evapotranspiration of area A-B affected by the ocean current?   (1) It is reduced.   (2) It is increased.   (3) It is unchanged.

TOPIC IX   Weathering and Erosion

WEATHERING

Weathering is the breakdown of rocks into particles called **sediment.** Rocks weather because they are adjusting to a new environment different from the one in which they were formed. Weathering occurs when rocks are exposed to the air, water, and living things at or near the earth's surface.

TYPES OF WEATHERING. There are two general types of weathering—*physical* and *chemical.*

Physical weathering is the breakdown of rock into smaller pieces (sediment) without chemical change. Physical weathering includes *frost action,* which breaks up rocks when water in the rocks freezes and expands.

Chemical weathering is the breakdown of rock by chemical action and results in a change in the mineral composition. Chemical weathering includes the uniting of minerals with oxygen *(oxidation)* or with water *(hydration),* and the dissolving of minerals by acids.

FACTORS AFFECTING THE RATE AND TYPE OF WEATHERING. Many variables determine the rate and type of weathering that will occur:

1. *Exposure.* The more exposed the rocks are to the air, water, and living things near the earth's surface, the faster the weathering.

2. *Particle size.* The smaller the particle size, the greater the total surface area per unit volume exposed to weathering and the greater the rate of weathering (see Figure 9-1).

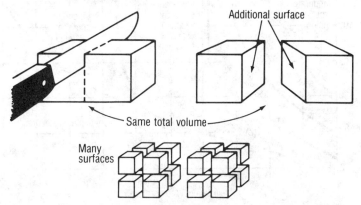

Figure 9-1. Explaining why surface area increases as particle size decreases. A single division of the block at the left exposes two new surface areas. If the two blocks at the bottom are divided again and again into smaller and smaller pieces, the total surface area will increase rapidly, although the total mass of material will remain the same.

3. *Mineral composition.* A rock's mineral composition determines the rate of weathering because different minerals have different resistances to weathering.

4. *Climate.* Climate greatly influences the rate and type of weathering (see Figure 9-2). Chemical weathering is most pronounced in warm, moist climates. The higher the average temperatures and the more humid the climate, the more intense the chemical weathering becomes. In cold climates, frost action (a type of physical weathering) is the most common and most effective form of weathering. It is especially intense where the climate is moist and the temperature variations lead to much alternate freezing and thawing.

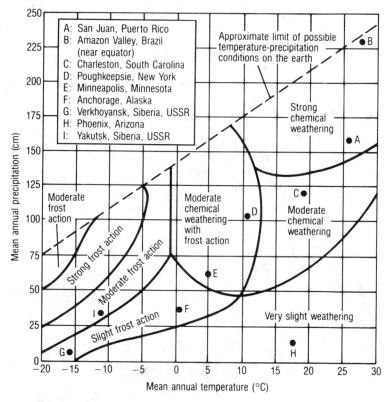

Figure 9-2. Dominant type of weathering for various climatic conditions. For example, if the mean annual temperature is 15°C and the mean annual precipitation is 100 cm, moderate chemical weathering is the main agent of weathering. The blank region of the chart in the upper left represents conditions that almost never occur on the earth.

PRODUCTS OF WEATHERING. Weathering results in the formation of three different types of sediment: (1) solid sediments such as sand and pebbles; (2) very small solid sediments called *colloids* (or clay-sized particles); and (3) dissolved ionic minerals, which are a major end product of much weathering. Dissolved minerals are what cause the "hardness" in all surface and ground water and what make the oceans "salty."

SOIL

One product of weathering is **soil,** which is the part of the regolith that will support rooted plants.

TYPES OF SOIL. Soil can be divided into two general types—*residual* and *transported.* A **residual soil** is a soil that formed from the rocks or regolith under the soil. A **transported soil** is one that has been moved into an area from another place by erosion.

SOIL FORMATION. The composition of soil varies with depth. This is indicated by horizontal layers or zones called **soil horizons,** which develop as a result of weathering processes. Soil and soil horizon development starts at the earth's surface and progresses downward with time, as illustrated in Figure 9-3.

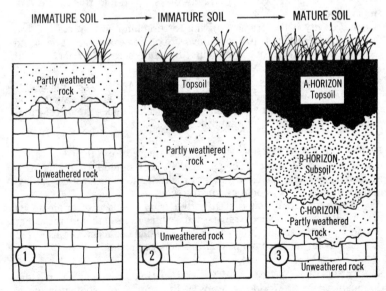

Figure 9-3. Profile development of horizons in a residual soil. Originally, there is unweathered bedrock at the surface. Weathering gradually proceeds downward to deeper levels. Meanwhile, infiltration, capillary migration, the addition of plant material, and the movement of sediment by organisms in the developing soil produce horizons with characteristic features.

Living things play a significant part in soil formation. For example, plant roots break up rock by widening cracks; accumulation of organic matter is an important part of soil; decay of organic matter forms acids that dissolve rock; earthworms, ants, and burrowing animals expose fresh rock material to weathering.

EROSION

Weathering is a process that produces most sediments; **erosion** is the process by which sediments are obtained and transported. Erosion is also the process by which the earth's surface is worn away, sculpted, and lowered.

EROSION THROUGH TRANSPORTING SYSTEMS.   Erosion of sediments is accomplished by means of *transporting systems.* A **transporting system** includes (1) the transporting agent (agent of erosion), such as running water (streams), glaciers, waves, density currents in water, wind, and people; (2) the "driving" force, which consists of gravity and the energy transfer from potential to kinetic energy; and (3) the material being transported, called sediment.

GRAVITY AND EROSION.   Gravity plays a two-fold role in the erosional process:

1. Gravity is the primary driving force behind all transporting systems, causing sediment to move downslope. For example, glaciers and streams move and erode sediments downhill because of the pull of gravity.

2. Gravity may act alone (that is, without the transporting agents mentioned above) in transporting earth materials. Loose sediment on steep slopes may break away and move downhill under the direct influence of gravity, such as in a landslide.

EVIDENCE OF EROSION.   Sediments displaced from their source are evidence of erosion. Such sediments can be seen in muddy streams, rock fields near glaciers, the assortment of rocks at shorelines, shifting sands, and earth moved by people in construction.

Weathered rock that is transported from its place of origin to its present location is called **transported sediment.** Weathered rock that remains in its place of origin is called **residual sediment.** Transported sediment is far more common than residual sediment.

STREAM, OR RUNNING WATER, EROSION.   Running water is the dominant agent of erosion in terms of the amount of sediment moved, even in many arid climates. Streams obtain sediments by (1) the direct lifting of particles from the stream bottom, or **stream bed;** (2) by collisions between carried sediments and the stream bed (abrasion); and (3) by receiving dissolved minerals, produced by weathering, from ground water flow or by directly dissolving the stream bed.

Streams carry sediments in three major ways: (1) dissolved minerals are carried in solution; (2) solid sediments of small size, including colloids, are carried by suspension; and (3) larger solid sediments are carried by rolling and bouncing along the bed (see Figure 9-4).

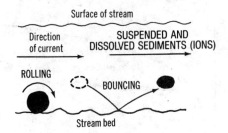

Figure 9-4. Transport of sediments by a stream.

STREAM VELOCITY.   The two most important factors that determine the average velocity of a stream are the *slope,* or *gradient,* of the stream and the *discharge,* or volume of water. Generally, as either the slope or the discharge of a stream increases, the velocity will increase. When a stream changes direction, the region of maximum velocity shifts as shown in Figure 9-5.

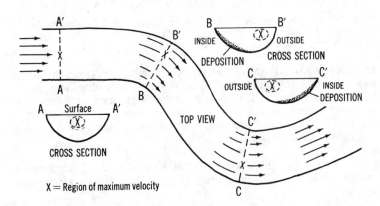

Figure 9-5. Variations in stream velocity in the cross section of an idealized stream. Where the stream course is straight (as at A), the location of maximum velocity is at the center of the cross section. Where the stream course changes direction (as at B and C), the location of maximum velocity moves toward the outside of the curve, or meander. Low velocity at the inside of curves results in deposition of the larger and denser sediments. Velocity also varies with depth. It is greatest just below the surface and least near the stream bed. At the air-water interface and at the stream bed velocity is reduced by friction.

SEDIMENT SIZE AND STREAM VELOCITY. The sediments being transported by a stream generally flow much slower than the stream. The greater the velocity of a stream, the larger the sediment particles it can carry (see Figure 9-6). For any given stream, the greater its velocity the more total sediment it can carry.

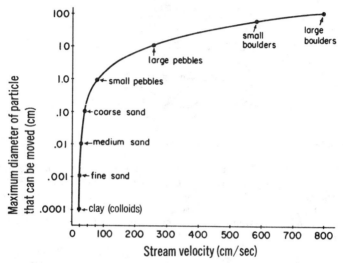

Figure 9-6. Relationship between stream velocity and maximum size of particle that can be moved. (Also see *Relationship of Transported Particle Size to Water Velocity* in the *Earth Science Reference Tables.*)

EROSION BY WIND AND GLACIERS. Many of the factors that affect stream erosion are found also in glacial and wind erosion. The greater the slope of a glacier, the faster it generally moves. The positions of maximum velocity in a glacier are similar to those of a stream shown in Figure 9-5. The greater the velocity of the wind, the larger the size of sediment particles it can carry.

SEDIMENT FEATURES AND EROSIONAL AGENT. Each agent of erosion produces distinctive characteristics in the sediment it transports.

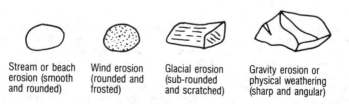

Figure 9-7. Sediment features reflect the erosional or weathering agents that produced or transported them.

A solid sediment in wind or stream erosion will become rounded and smaller. The longer the time the sediment is eroded, the more smooth and rounded it becomes. Wind-blown sediments are often more frosted (pitted) than stream sediments. Glacial sediments are only partly rounded and often have scratches of various sizes. Sediments produced by physical weathering or moved by gravity are often very angular in shape.

PEOPLE AND EROSION.   People greatly increase the erosion of land by their activities. The burning and cutting down of forests, poor farming methods, road and building construction, and mining expose loose soil and regolith to the agents of erosion and thereby increase erosion rates. The removal and lowering of the soil, regolith, and vegetation at the earth's surface is an aspect of erosion called **denudation.** Denudation has been greatly increased in the last few hundred years by the activities of people.

VOCABULARY

weathering	soil horizon
sediment	erosion
physical weathering	transporting system
chemical weathering	transported sediment
soil	residual sediment
residual soil	stream bed
transported soil	denudation

QUESTIONS ON TOPIC IX—WEATHERING AND EROSION

Questions in Recent Regents Exams (end of book)

June 1995:   28, 31, 32, 81–83, 85
June 1996:   30–33, 103
June 1997:   2, 26, 29–31, 39, 81–85, 104
June 1998:   28, 29, 31, 56, 58

Questions from Earlier Regents Exams

1. What is the main difference between chemical and physical weathering?   (1) Chemical weathering alters the composition of minerals, and physical weathering does not.   (2) Chemical weathering increases the surface area of minerals, and physical weathering does not. (3) Physical weathering alters the composition of minerals, and chemical weathering does not.   (4) Physical weathering increases the surface area of minerals, and chemical weathering does not.

2. Which property of water makes frost action a common and effective form of weathering?   (1) Water dissolves many earth materials.   (2) Water expands when it freezes.   (3) Water cools the surroundings when it evaporates.   (4) Water loses 80 calories of heat per gram when it freezes.

3. Chemical weathering will occur most rapidly when rocks are exposed to the   (1) hydrosphere and lithosphere   (2) mesosphere and thermosphere   (3) hydrosphere and atmosphere   (4) lithosphere and atmosphere

4. Which is the best example of physical weathering?   (1) the cracking of rock caused by the freezing and thawing of water   (2) the transportation of sediment in a stream   (3) the reaction of limestone with acid rainwater   (4) the formation of a sandbar along the side of a stream

5. The diagram below represents a geologic cross section with sedimentary rock layers *A*, *B*, *C*, and *D* exposed to the atmosphere. Which rock layer in the diagram is most resistant to weathering and erosion?   (1) *A*   (2) *B*   (3) *C*   (4) *D*

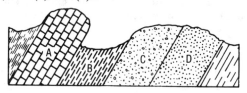

6. A rock normally weathers faster when broken up because of an increase in the rock's   (1) mass   (2) volume   (3) density   (4) total surface area

7. In which climate would the chemical weathering of limestone occur most rapidly?   (1) cold and dry   (2) cold and humid   (3) warm and dry   (4) warm and humid

8. In which climate does physical weathering by frost action most easily occur?   (1) dry and hot   (2) dry and cold   (3) moist and hot   (4) moist and cold

9. Two different kinds of minerals, *A* and *B*, were placed in the same container and shaken for 15 minutes. The diagrams below represent the size and shape of the various pieces of mineral before and after shaking. What caused the resulting differences in shapes and sizes of the minerals?

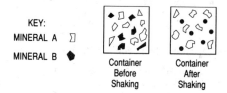

KEY:

MINERAL A

MINERAL B

Container Before Shaking

Container After Shaking

(1) Mineral *B* was shaken harder.   (2) Mineral *B* had a glossy luster.   (3) Mineral *A* was more resistant to abrasion.   (4) Mineral *A* consisted of smaller pieces before shaking began.

10. Which is indicated by a deep residual soil?   (1) resistant bedrock   (2) a large amount of glaciation   (3) a long period of weathering   (4) a youthful stage of erosion

11. The chemical composition of a residual soil in a certain area is determined by the   (1) method by which the soil was transported to the area   (2) slope of the land and the particle size of the soil   (3) length of time since the last crustal movement in the area occurred (4) minerals in the bedrock beneath the soil and the climate of the area

12. A variety of soil types are found in New York State primarily because areas of the state differ in their (1) amounts of insolation (2) distances from the ocean (3) underlying bedrock and sediments (4) amounts of human activities

13. Soil horizons develop as a result of (1) capillary action and solution (2) erosion and ionization (3) leaching and color changes (4) weathering processes and biologic activity

14. Particles of soil often differ greatly from the underlying bedrock in color, mineral composition, and organic content. Which conclusion about these soil particles is best made from this evidence? (1) They are residual sediments. (2) They are transported sediments. (3) They are uniformly large-grained. (4) They are soluble in water.

15. Which change would cause the topsoil in New York State to increase in thickness? (1) an increase in slope (2) an increase in biologic activity (3) a decrease in rainfall (4) a decrease in air temperature

16. The primary force responsible for most of the transportation of rock material on the surface of the earth is (1) gravity (2) wind (3) running water (4) glaciers

17. Which erosional force acts alone to produce avalanches and landslides? (1) gravity (2) winds (3) running water (4) sea waves

18. Transported rock materials are more common than residual rock materials in the soils of New York State. Which statement best explains this observation? (1) Solid rock must be transported to break. (2) Weathering changes transported rock materials more easily than residual rock materials. (3) Most rock materials are moved by some agent of erosion at some time in their history. (4) Residual rock material forms only from bedrock that is difficult to change into soil.

19. The composition of sediments on the earth's surface usually is quite different from the composition of the underlying bedrock. This observation suggest that most (1) bedrock is formed from sediments (2) bedrock is resistant to weathering (3) sediments are residual (4) sediments are transported

20. Which is the best evidence that erosion has occurred? (1) a soil rich in lime on top of a limestone bedrock (2) a layer of basalt found on the floor of the ocean (3) sediments found in a sandbar of a river (4) a large number of fossils embedded in limestone

21. Which agent probably contributes most to the general wearing down of the earth's surface? (1) wind (2) glaciers (3) running water (4) ocean waves

22. The increase of dissolved materials in the ocean is primarily the result of the (1) abrasion of the ocean floor (2) transporting of material by rivers (3) melting of continental glaciers (4) deposition of materials by ground water

23. How are dissolved particles of sediment carried in a river? (1) by bouncing and rolling   (2) by precipitation   (3) in solution (4) in suspension

24. According to the *Earth Science Reference Tables*, which material would most easily be carried in suspension by a slow-moving stream? (1) clay   (2) silt   (3) sand   (4) gravel

25. In summer, a small stream has a depth of 3 meters and a velocity of 0.5 meter per second. In spring, the same stream has a depth of 5 meters. The velocity of the stream in spring is more likely closest to (1) 0.1 m/sec   (2) 0.2 m/sec   (3) 0.5 m/sec   (4) 0.8 m/sec

26. The velocity of a stream at a particular location is controlled mainly by the   (1) elevation of the stream at the location   (2) distance of the location from the source   (3) slope of the stream bed at the location   (4) amount of sediments carried at the location

27. As the kinetic energy of a stream increases, the size of the particle that can be moved   (1) decreases   (2) increases   (3) remains the same

28. A pebble is being transported in a stream by rolling. How does the velocity of the pebble compare to the velocity of the stream? (1) The pebble is moving slower than the stream.   (2) The pebble is moving faster than the stream.   (3) The pebble is moving at the same velocity as the stream.

29. Which graph best represents the relationship between the maximum particle size that can be carried by a stream and the velocity of the stream?

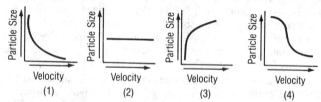

(1)   (2)   (3)   (4)

30. According to the *Earth Science Reference Tables*, which is the largest sediment that could be carried by a stream flowing at a velocity of 75 centimeters per second?   (1) silt   (2) sand   (3) pebbles   (4) cobbles

31. According to the *Earth Science Reference Tables*, which is the *slowest* stream velocity needed to maintain 1 cm particles moving downstream?   (1) 50 cm/sec   (2) 75 cm/sec   (3) 100 cm/sec   (4) 125 cm/sec

32. Which landscape characteristic best indicates the action of glaciers?   (1) few lakes   (2) deposits of well-sorted sediments   (3) residual soil covering large areas   (4) polished and scratched surface bedrock

33. While studying the movement of valley glaciers that are advancing from the north, geologists placed metal stakes extending in a straight line from one valley wall to the other. Which sketch would best illustrate the position of the stakes one year later?

34. Compared to its original shape, a rock that has been transported several kilometers by a stream will normally be more   (1) rounded (2) jagged   (3) flattened on one side   (4) rectangular

35. Based on the diagrams of rock fragments below, which shows the *least* evidence of erosion?

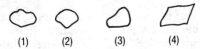

36. Sharp-edged, irregularly shaped sediment particles found at the base of a rock cliff were probably transported by   (1) gravity   (2) wind (3) ocean waves   (4) running water

37. As the amount of plant life in a region is decreased by the activities of humans, the amount of erosion will probably   (1) decrease (2) increase   (3) remain the same

38. The graph below compares erosion caused by changes in the discharge of a stream before and after a period of population growth.

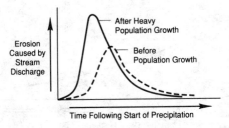

The best interpretation of the graph is that the total amount of erosion (1) decreases with an increase in population   (2) increases with an increase in population   (3) is not affected by an increase in population

TOPIC X Deposition

Deposition, also called *sedimentation,* is the process by which sediments are released, settled from, or dropped from an erosional system. Deposition includes the releasing of solid sediments and the process of *precipitation,* which is the releasing of dissolved ionic minerals from a water solution. Most final deposition occurs in large water bodies, because running water is the dominant erosional system.

FACTORS CAUSING DEPOSITION

Deposition usually occurs when the velocity of the stream, wind, or other erosional system decreases. Some of the factors that determine which sediments are deposited, and their rates of settling, are described below.

1. Size. All other factors being equal, when wind or running water slows down, the larger sediments settle out first. This occurs because the larger sediment particles are heavier and therefore sink faster. Very small particles (less than 0.0004 cm in diameter), called **colloids** (or clay), may remain suspended in water almost indefinitely.

2. Shape. The shape of a particle helps determine how fast it will be deposited from wind or running water. All other factors being equal, the more spherical a sediment, the faster it will settle out, and the more flattened it is, the greater its resistance to deposition.

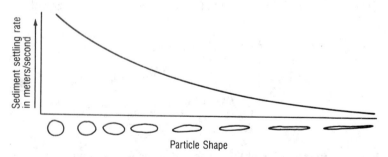

Figure 10-1. Settling rates of sediments of various shapes. As sediments of equal volume and density become less spherical (or rounded) and more flattened it takes more time for them to settle, so settling rate decreases (gets slower).

3. Density. All other factors being the same, the higher the density of a sediment, the faster it will settle out of air or water. The reason for this is that if two particles have the same size and shape, the denser one will be heavier.

138

SORTING OF SEDIMENTS AND DEPOSITION

Sorting of sediments is the degree of similarity of sediment size in a given deposit. The greater the similarity of size, the more sorted the sediments are said to be.

GRADED BEDDING.   When a mixture of sediment sizes in water settles out rapidly, a horizontal layer *(bed)* develops with sediment size decreasing toward the top of the bed. Such an arrangement in a sediment layer is called **graded bedding** (see Figure 10-2). Graded bedding is most often associated with sediment-laden density currents, called *turbidity currents*. These turbidity currents are most common on the sloped ocean bottoms off the coasts of continents, and produce graded bedding on the flatter ocean bottom further offshore *(abyssal areas)*. Graded bedding can also form in lakes fed by streams.

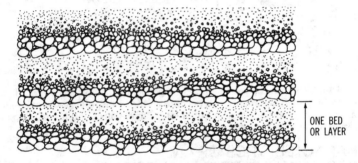

ONE BED
OR LAYER

Figure 10-2. Layers of sediment with graded bedding. In each bed or layer, sediment sizes decrease from bottom to top. Graded bedding results from rapid deposition in an erosional system, as by a turbidity current (see text). Each bed is the result of a single turbidity flow.

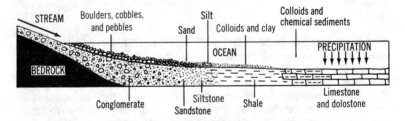

Figure 10-3. Horizontal sorting where a stream enters the ocean. The larger sediments settle first (nearer the shore). Sediments become smaller as distance from shore increases. Where solid sediments are rare, chemical precipitation is the dominant type of deposition. Precipitation nearer the shore may provide the cement for the formation of the types of sedimentary rocks indicated. Limestone and dolostone are common evaporite sedimentary rocks formed largely by precipitation.

HORIZONTAL SORTING.   When the velocity of a wind or water erosional system gradually decreases (such as when a stream flows into the
ocean at a delta), the larger, denser, and more spherical sediments settle
out first. This results in layers with **horizontal sorting,** in which the
sediment size, sphericalness, and density generally decrease in the direction the erosional system was moving (see Figure 10-3).

UNSORTED GLACIAL DEPOSITS.   In a solid erosional system such
as a glacier, sediments of all sizes, shapes, and densities are deposited
together. This results in the unsorted deposits characteristic of direct
glacial deposits (see Figure 10-4).

Figure 10-4. Unsorted glacial deposits. In glacial deposition there is no fluid
medium in which sediments can become sorted. Thus, glacial deposits are
characterized by a random distribution of sediment sizes and no bedding or
layering. Note glacial scratches on sediments.

CHARACTERISTICS OF AN
EROSIONAL-DEPOSITIONAL SYSTEM

MODEL OF AN EROSIONAL-DEPOSITIONAL SYSTEM.   Figure 10-
5 shows a side view and a top view of an imaginary stream used as a
model of an erosional-depositional system. In this simple model it is
assumed that the volume of flow (discharge) is the same throughout the
length of the stream.

ENERGY TRANSFORMATIONS IN THE MODEL SYSTEM.   At the
beginning, or *source,* of the stream (A), the system has a maximum of
potential energy. As the stream flows toward its end, or *mouth* (D),
potential energy is continuously being transformed into kinetic energy.
This kinetic energy is at the same time being lost to the environment in
the form of heat produced by friction. Where the slope of the stream is
steep, the transformation of energy occurs most rapidly, the stream has
its greatest velocity, and the system has its greatest kinetic energy. Where
the slope is small, the rate of energy transformation decreases, the stream
slows down, and the kinetic energy of the system decreases. At the
mouth of the stream, the velocity drops to zero, and the system has zero
kinetic energy. Since the system has less potential energy because of its
lower elevation, there has been a net loss of energy between the source
and the mouth.

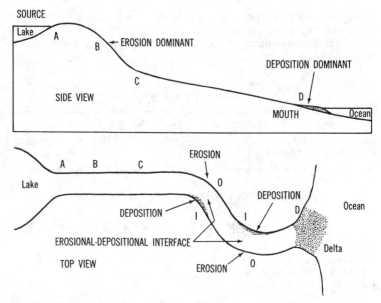

Figure 10-5. Model of a stream representing an erosional-depositional system. "O" indicates the outside of a curve and "I" indicates the inside. Some deposition may occur at **C**, where the slope and velocity decrease. Erosional-depositional interfaces will then exist between **B** and **C** and between **C** and **D**.

EROSION AND DEPOSITION IN RELATION TO ENERGY CHANGES. Wherever the kinetic energy of the system is large, erosion is the dominant process. Where the kinetic energy is small, deposition is the dominant process. Thus, erosion occurs in regions of steep slope, and deposition occurs in regions of gentle slope. Deposition is particularly rapid at the mouth of the stream, where the kinetic energy becomes zero. Stream velocity is also larger at the outside of curves, or meanders, and smaller at the inside. Therefore, erosion usually occurs at the outside of curves and deposition occurs at the inside.

EROSIONAL-DEPOSITIONAL INTERFACES. Since there are regions of erosion and regions of deposition along the length of the stream, an interface between an erosional and a depositional state can often be located in the system. Interfaces between erosion and deposition exist at the curves in the model stream, and near the mouth of the stream. They may also be found where changes in slope occur, as between B and C in Figure 10-5.

DYNAMIC EQUILIBRIUM OF THE SYSTEM. Since all sediments picked up by the stream during erosion must eventually be deposited, the system is in a state of *dynamic equilibrium*. Although erosion and deposition are occurring continuously, the rate of erosion equals the rate of deposition by the system as a whole. If a flood occurs, a stream will erode (pick up and transport) more sediment, but it will also deposit an equally increased volume of sediment, thus establishing a new balance of equilibrium.

VOCABULARY	
deposition	graded bedding
colloid (clay)	horizontal sorting
sorting of sediments	

QUESTIONS ON TOPIC X—DEPOSITION

Questions in Recent Regents Exams (end of book)

June 1995: 29, 33
June 1996: 30, 32, 33, 103
June 1997: 26, 30, 31, 84, 85
June 1998: 31, 32, 34, 76–80

Questions from Earlier Regents Exams

1. The rate at which particles are deposited by a stream is *least* affected by the   (1) size and shape of the particles   (2) velocity of the steam   (3) stream's elevation above sea level   (4) density of the particles

2. The graph below shows the relationship between particle shape and settling rate.

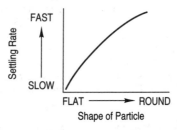

Which statement best describes the relationship shown?   (1) Flatter particles settle more slowly than rounder particles.   (2) Flatter particles settle faster than rounder particles.   (3) All particles settle at the same speed.   (4) Particle shape does not affect settling rate.

3. A mixture of sand, pebbles, clay, and silt, of uniform shape and density, is dropped from a boat into a calm lake. Which material most likely would reach the bottom of the lake first?   (1) sand   (2) pebbles   (3) clay   (4) silt

4. A low hill is composed of unsorted sediments that have mixed grain sizes. This hill was probably deposited by   (1) a glacier   (2) the wind   (3) running water   (4) wave action

5. Small spheres that are identical in shape and size are composed of one of four different kinds of substances: *A, B, C,* or *D.* The spheres are mixed together and poured into a clear plastic tube filled with water. Which property of the spheres caused them to settle in the tube as shown in the diagram below?   (1) their size   (2) their shape   (3) their density   (4) their hardness

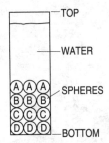

6. A stream is entering the calm waters of a large lake. Which diagram best illustrates the pattern of sediments being deposited in the lake from the stream flow?

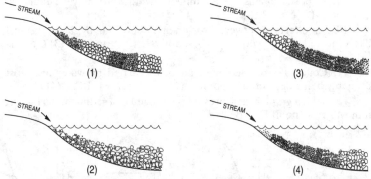

7. The map below represents a winding stream. At which location is stream deposition most likely to be greater than stream erosion?   (1) *A*   (2) *B*   (3) *C*   (4) *D*

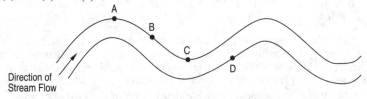

8. If the gradient of a stream increases at a certain location, the probability that sediment will be deposited at that location   (1) decreases   (2) increases   (3) remains the same

Base your answers to questions 9 through 12 on the *Earth Science Reference Tables* and the diagram below. The diagram represents a glacier moving out of a mountain valley. The water from the melting glacier is flowing into a lake. Letters *A* through *F* identify points within the erosional/depositional system.

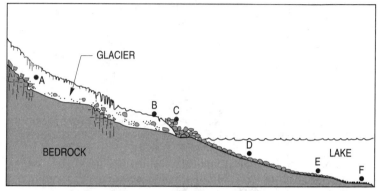

9. Deposits of unsorted sediments would probably be found at location   (1) *E*   (2) *F*   (3) *C*   (4) *D*

10. An interface between erosion and deposition by the ice is most likely located between points   (1) *A* and *B*   (2) *B* and *C*   (3) *C* and *D*   (4) *D* and *E*

11. Colloidal-sized sediment particles carried by water are most probably being deposited at point   (1) *F*   (2) *B*   (3) *C*   (4) *D*

12. Which graph best represents the speed of a sediment particle as it moves from point *D* to point *F*?

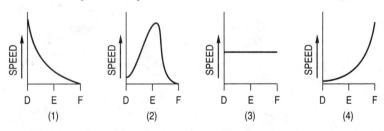

13. More deposition than erosion will take place in a streambed when the   (1) density of the rock particles carried by the stream decreases   (2) slope of the stream increases   (3) discharge of the stream increases   (4) velocity of the stream decreases

14. Which is the most probable description of the energy of a particle in an erosional-depositional system?   (1) Particles gain kinetic energy during erosion and lose kinetic energy during deposition. (2) Particles lose kinetic energy during erosion and lose kinetic energy during deposition.   (3) Particles gain potential energy during erosion

and gain potential energy during deposition.   (4) Particles lose potential energy during erosion and gain potential energy during deposition.

Additional Questions

To answer questions 1 through 8 refer to the diagram below, which shows
the top and side views of a stream. Consider the volume of water in the
stream to be everywhere the same.

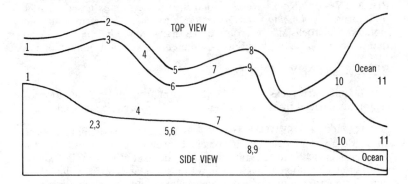

   1. At which place in the stream would there be the most clear interface between erosion and deposition?   (1) 1   (2) 2-3   (3) 3-4   (4) 7
   2. What is true about the total amount of energy of this stream as
it flows from 1 to 10?   (1) It steadily increases.   (2) It steadily decreases.   (3) It increases and decreases.   (4) It remains the same.
   3. Between what two points would the most potential energy be
converted to kinetic energy?   (1) 1-2   (2) 2-5   (3) 5-8   (4) 8-10
   4. At which location would the most deposition occur because of
loss of energy?   (1) 4   (2) 6   (3) 9   (4) 10
   5. At which location would the stream be doing the most eroding of
its side and banks?   (1) 1   (2) 3   (3) 5   (4) 8
   6. If the stream is eroding as much as it is depositing between 8
and 10, the stream at this location can be said to   (1) be an interface
(2) have dynamic equilibrium   (3) be sorted   (4) have equal amounts
of kinetic and potential energy
   7. Which characteristic would usually decrease the most between
locations 10 and 11?   (1) the amount of salt in solution   (2) the size of
the sediments   (3) the density of the water   (4) the depth of the water
   8. As the water of the stream flows from 1 to 10, the total potential
energy of the stream will   (1) increase   (2) decrease   (3) remain the
same

TOPIC XI   Rock Formation

Rocks are any naturally formed solid matter that makes up a part of the solid earth. Rocks are usually composed of a mixture of minerals. The **texture** of a rock is the size, shape, and arrangement of the grains (mineral crystals or sediments) of which it is composed.

ROCK TYPES. Rocks are classified into three groups based on how they formed: *sedimentary, metamorphic,* and *igneous.* The igneous and metamorphic rocks may be grouped together as *nonsedimentary rocks,* so that rocks may also be classified as sedimentary and nonsedimentary.

SEDIMENTARY ROCKS

Sedimentary rocks are rocks that form from an accumulation of sediments derived from preexisting rocks and/or organic material.

FORMATION OF SEDIMENTARY ROCKS. Most sedimentary rocks are made up of solid sediments, or **clasts,** weathered from other rocks. The sediments are then transported and deposited by running water, wind, or glaciers to new locations in water or on land. Most sedimentary rocks form under bodies of water such as lakes and oceans. Sedimentary rocks are formed in the following ways:

    1. Cementation. Larger solid sediments, such as sand and pebbles, form rocks when the sediments are cemented together by minerals that are precipitated out of solution (see Figure 11-1, A and B).

Figure 11-1. Some characteristics of sedimentary rocks. (A) Conglomerate. Particles of unsorted sizes cemented together. **(B)** Sandstone. Particles of fairly uniform (sorted) size cemented together. **(C)** Shale. Rock formed by compression of very small (colloidal) sediments, possibly cemented as well. **(D)** Chemical sedimentary rock. Monomineralic crystalline rock (composed of intergrown mineral crystals) formed by evaporation and/or precipitation.

    2. Compression/Compaction. When very small solid sediments (colloids) are compressed by the pressure of overlying water and sediments, they form rocks like the one in Figure 11-1C. Sedimentary rocks may also be formed by a combination of compression and cementation.

3. Chemical action. Ionic or dissolved minerals may **precipitate** out as the result of chemical processes or of evaporation of the water. A rock formed in this manner is **monomineralic** rock, that is, it consists of **crystals,** or grains, of a single mineral (see Figure 11-1D). Rocks formed by these chemical processes are called **chemical sedimentary rocks** or evaporites. Chemical limestone and rock salt are evaporites.

4. Biologic processes.   Some sedimentary rocks form as a result of biologic processes. Many forms of life that live in water, such as clams and corals, extract dissolved minerals from the water to form their hard parts. When the organisms die, these hard parts accumulate to form rocks such as fossil limestone. Bituminous coal is a rock formed when dead plant matter is deposited in water and compressed.

CHARACTERISTICS OF SEDIMENTARY ROCKS.

1. Most sedimentary rocks are clastic, composed of fragments or particles which strongly resemble sediments.

2. Some sedimentary rocks have a range of particle or sediment sizes (see Figure 11-1A).

3. Other sedimentary rocks consists mainly of one sediment size because of the sorting of sediments (see Figures 11-1B and 11-C).

4. Some sedimentary rocks are of **organic** origin; that is, they are composed of plant and animal products or remains.

5. Sedimentary rocks often have parallel layers, as shown in Figure 11-2. Such layers are also called **beds** or **strata.**

6. Sedimentary rocks often contain *fossils.*

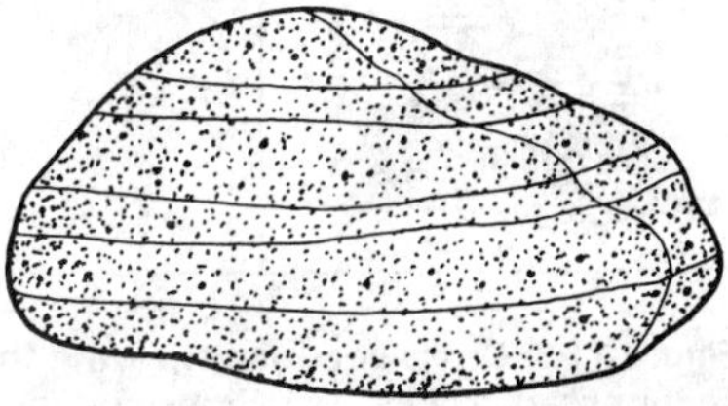

Figure 11-2. Layers in a sedimentary rock. Layers may result from changes in the type of sediment during the period of deposition.

IDENTIFYING SEDIMENTARY ROCKS. Refer to the chart *Scheme for Sedimentary Rock Identification* in the *Earth Science Reference Tables.* As can be inferred from the chart, sedimentary rocks are classified and identified largely on the basis of size of solid sediments (particle diameter) and/or mineral composition of the rock. Monomineralic chemical sedimentary rocks, such as chemical limestone and dolostone, are identified primarily based on the one mineral they contain. For example, rock gypsum is composed of the mineral gypsum, and dolostone is composed of dolomite.

The vast majority of sedimentary rocks are composed of solid sediments (boulders to clay) that have been cemented and/or compacted together. These clastic sedimentary rocks are most commonly identified according to the size of the majority of the sediments they contain, regardless of their mineral composition. For example, if a rock is composed largely of solid sediments of 0.01 cm in diameter (sand size), the rock would be called a sandstone.

NONSEDIMENTARY ROCKS

Nonsedimentary rocks are those rocks that do not form from sediments. There are two groups of nonsedimentary rocks—the *igneous* and the *metamorphic* rocks.

IGNEOUS NONSEDIMENTARY ROCKS. **Igneous rocks** are those rocks that form from the cooling and **solidification** of liquid rock. (Liquid rock beneath the earth's surface is known as *magma*. When liquid rock reaches the earth's surface, it is called *lava*.) When liquid rock cools, it solidifies, usually by the growth of various mineral crystals in a type of solidification called **crystallization**. This results in a **crystalline** texture, consisting of intergrown mineral crystals of different sizes, shapes, and chemical composition, as shown in Figure 11-3. Such a rock is called **polymineralic**, because it contains two or more different minerals. Solidification includes the change from liquid rock to a noncrystalline glassy igneous rock (as shown in Figure 11-3C) and it includes the process of crystallization.

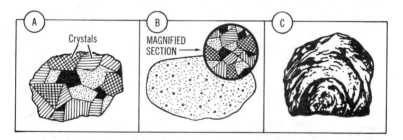

Figure 11-3. Cross sections showing three types of texture in igneous nonsedimentary rocks. (A) Slow cooling of the liquid rock results in large crystals visible to the unaided eye (coarse texture). **(B)** If cooling is more rapid, the crystals are too small to be seen without magnification (fine texture). A magnified section resembles the texture in **A.** **(C)** If cooling is very rapid, the solid rock has a glassy texture without mineral crystals.

1. Size of crystals. The size of the crystals in an igneous nonsedimentary rock depends on the conditions of the environment in which the rock formed. The immediate cause of the difference in the size of the crystals is the amount of time in which the cooling takes place. The longer the time of cooling, the larger the crystals become. However, the cooling time itself depends on the temperature and pressure of the environment. The pressure and temperature deep within the earth are very high, and therefore liquid rock cools slowly (possibly for thousands of years). The result is rocks with large crystals. (A body of rock formed by solidification below the earth's surface is called an **intrusion,** and the igneous rocks that form are called **intrusive** or **plutonic.**)

The temperature and pressure at or near the earth's surface are much lower, and liquid rock cools there much more quickly than at great depths. This produces rocks with much smaller crystals. (A body of rock formed by solidification at the earth's surface is called an **extrusion,** and the igneous rocks that form are called **extrusive** or **volcanic.**)

2. Texture of igneous rocks. The texture of igneous rocks depends on the size of the crystals, as well as on their shape and arrangement within the rock. Rocks with large crystals have a coarse texture (see Figure 11-3A). Those with crystals that are too small to be seen clearly with the unaided eye have a fine texture. Igneous rocks that cool very rapidly at the earth's surface have no crystals at all, and have a glassy texture. Some extrusive igneous rocks have a porous texture resulting from solidification around gas bubbles.

IDENTIFYING IGNEOUS ROCKS. Refer to the *Scheme for Igneous Rock Classification* in the *Earth Science Reference Tables*. The diagram shows a common method of classifying and identifying igneous rocks. Igneous rocks are identified largely on the basis of texture (coarse, fine, glassy, or porous) and percent mineral composition. The majority of the mineral crystals in a coarse-grained rock are easily visible to the unaided eye (1 mm or larger), while the majority of the mineral crystals in a fine-grained rock are *not* easily seen by the unaided eye.

Percent mineral composition divides igneous rocks on the diagram into vertical columns (igneous rock families). Within a rock family, rocks have similar mineral compositions.

Other information indicated by the *Scheme for Igneous Rock Identification* includes:

1. As rocks appear farther to the right on the diagram, they become more mafic (high in magnesium and iron) and less felsic or granitic (high in silicon and aluminum).

2. As rocks appear farther to the right on the diagram, their density increases.

3. As rocks appear farther to the right on the diagram, their overall color becomes darker.

METAMORPHIC NONSEDIMENTARY ROCKS. **Metamorphic rocks** are those rocks that are formed from other rocks (igneous, sedimentary, or other metamorphic rocks) within the lithosphere. The formation of metamorphic rocks occurs in response to conditions of heat, pressure, and/or chemical change. Such conditions are often associated with the pressure, heat, and deep burial that result from mountain building processes; therefore, metamorphic rocks are often found in mountainous regions where weathering and erosion have exposed rock that was once deeply buried. **Regional metamorphism** occurs when metamorphic rocks occur in large areas due to mountain-building processes such as folding and faulting. Under conditions of high temperature and pressure, many metamorphic rocks form by the process of **recrystallization**, which is a growth of mineral crystals at the expense of surrounding sediments or other crystals of the original rock (see Figure 11-4). Recrystallization occurs without true melting.

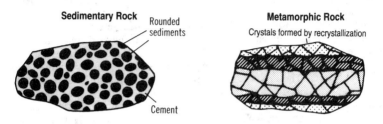

Figure 11-4. Metamorphic rock formed by recrystallization. Under the influence of heat and pressure, minerals in the sediments or cement of a sedimentary rock combine to form mineral crystals, producing a crystalline texture. However, recrystallization occurs without melting of the rock.

Metamorphic nonsedimentary rocks often have certain textural and structural characteristics that result from the environment of formation:

1. Foliation is a layered arrangement (texture) in a metamorphic rock in which firmly joined crystals of like minerals are aligned in layers (see Figure 11-4). The layers are formed when the rock is subjected to extreme environmental pressure and high temperatures. Usually, the greater the pressure and temperature, the thicker the foliations (see Figure 11-5).

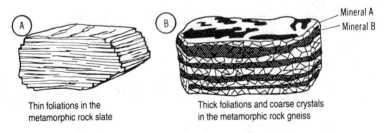

Figure 11-5. Foliation in metamorphic nonsedimentary rocks.

2. Distorted structure is the curving and folding of the foliations in metamorphic rocks (see Figure 11-6). These distortions of the once-horizontal layers are caused by great environmental pressures exerted on the rocks from different directions.

ENVIRONMENT OF ROCK FORMATION

The type of environment in which a rock formed is inferred from its composition, structure, and texture. The following are examples of such inferences.

1. Composition. If a sedimentary rock is made of halite, the rock probably formed by means of evaporation and precipitation in a body of salt water. The presence of fossils indicates formation at or near the earth's surface.

2. Structure. The distorted structure of the metamorphic rock shown in Figure 11-6 was caused by an environment of great pressure.

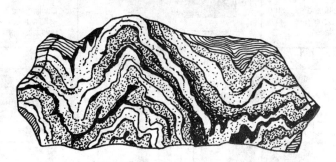

Figure 11-6. Distorted structure in a metamorphic nonsedimentary rock.

3. Texture. The texture of a rock results from the size, shape, and arrangement of its mineral crystals or sediments. If the sediments found in a sedimentary rock are angular in shape, it may be inferred that the rock was formed near the source of the sediments.

DISTRIBUTION OF ROCK TYPES.   Sedimentary rocks are usually found as a *veneer* (relatively thin coating) over large areas of continents. The nonsedimentary rocks that occur at or near the surface are most often found in regions of volcanoes and mountains (see Figure 11-7). The nonsedimentary rocks in mountain regions are exposed on the surface after millions of years of weathering and erosion have removed the veneer of sedimentary rocks.

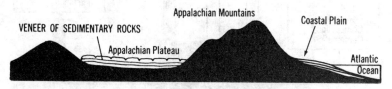

Figure 11-7. Veneer of sedimentary rocks in the Appalachian Plateau in the eastern United States. The solid black indicates nonsedimentary rock.

THE ROCK CYCLE.   The **rock cycle** is a model used to show how the rock types (sedimentary, igneous, and metamorphic) are interrelated, and the processes that produce each rock type. Figure 11-8 on page 152 is a diagram of the rock cycle.

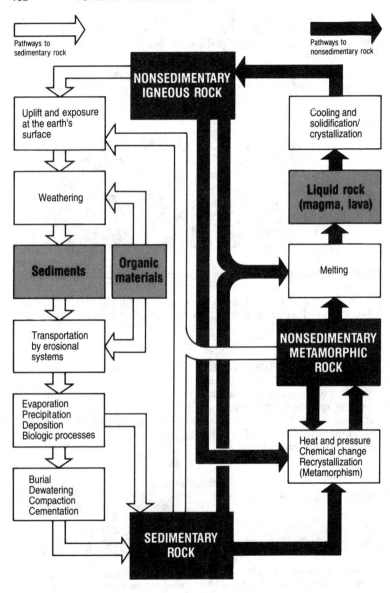

Figure 11-8. The rock cycle. Some of the processes by which one type of rock can be changed to another.

Some of the major concepts of the rock cycle are:

1. *Any one rock type can change into any other rock type.* Thus a specimen of any one type can have materials in it that were once part of any other type.

2. *There is no preferred direction of movement of materials in the rock cycle for any one mass of material.* Any one piece of material can stay in any one place for any length of time, or it can follow any of the arrows indicated on the diagram.

3. *There is no exact point of separation between the rock types.* For example, where local rock has been in contact with an intrusion of molten rock, or with a lava flow (extrusion), there is a **transition zone (contact metamorphic zone)** between the original local rock and the intrusion or the lava flow. In this transition zone there is a blending of the rock types and often no exact boundary (see Figure 11-9).

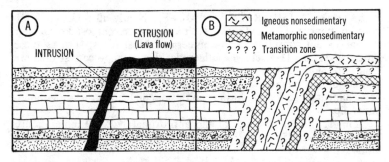

Figure 11-9. Formation of transition zones between rock types by a flow of molten rock. In **A**, molten rock has flowed up through a crack in sedimentary rock to the surface, forming an intrusion below the surface and an extrusion (lava flow) on the surface. In the contact zone between the original local rock and the intrusion or extrusion, there is a blending of rock types from sedimentary, through metamorphic nonsedimentary, to igneous nonsedimentary.

4. *Sedimentary rocks often contain sediments (clasts) that have varied origins.* This occurs because (a) sedimentary rocks form from the weathered products of any type of rock and/or organic materials, and (b) in any one depositional basin where sedimentary rocks form, the sediments can be brought from many different areas and by various methods of transport.

5. *The composition of some rocks suggests that the materials (sediments or minerals) in the rock have undergone multiple transformations (changes) within the rock cycle.* This is so because of the many possible paths within the rock cycle that can produce any one rock. For example, a sedimentary rock may contain fragments, or sediments, that were once part of a metamorphic rock that formed deep within the earth. However, that metamorphic rock itself may have been transformed from a sedimentary rock that formed near the earth's surface. A sedimentary rock may also contain sediments from previous sedimentary rocks and fragments which were once part of an intrusive igneous rock. The cement holding the rock together may once have been part of an igneous rock that was uplifted to the earth's surface and chemically weathered to form dissolved minerals; the minerals were then transported by water and later precipitated to become the cement.

MINERALS

A **mineral** is a naturally occurring, crystalline, inorganic substance with characteristic physical and chemical properties. That is, a mineral is a natural substance, a solid with a specific arrangement of constituent units, and either a chemical element or compound, which is not derived from parts of organisms (forms of life) or their products.

RELATION OF MINERALS TO ROCKS. Nearly all rocks are composed of minerals. (A few rocks, such as coal, are composed of organic substances not considered to be minerals.) Most rocks, such as those shown in Figure 11-5, are **polymineralic,** that is, they consist of more than one mineral. Some rocks are **monomineralic,** that is, they consist entirely of one mineral. Many sedimentary rocks and the metamorphic nonsedimentary rocks that form from those sedimentary rocks are monomineralic.

Only 20 to 30 minerals of the approximately 2,500 minerals are commonly found in most rocks. These are called the **rock-forming minerals.**

MINERAL COMPOSITION. Minerals are composed of *elements.* A few minerals, such as copper, sulfur, and graphite, are each composed of only one element. Most minerals are composed of two or more elements in chemical combination and are called *compounds.*

The circle graph and the table in Figure 11-10 illustrate the following points:

1. Most of the earth's crust and thus most minerals are made up of only a few elements.

2. Oxygen is the most abundant element by weight and volume.

3. Silicon is the second most abundant element by weight.

MINERAL PROPERTIES. Minerals have characteristic physical and chemical properties. Some of these properties are color, crystal shape, hardness, luster, streak, how the mineral breaks (cleavage and fracture), and density. Many of the physical properties of minerals are related to the structural arrangement of the mineral's constituent units (atoms or ions). A good example of this is the way the mineral halite (common salt) breaks into cubes because of a cubic arrangement of the constituent units (ions), as shown in Figure 11-11. Another example is the characteristic external crystal shape of minerals.

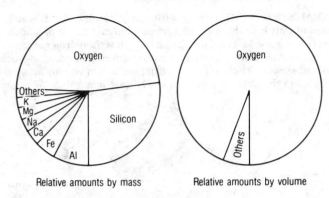

Relative amounts by mass   Relative amounts by volume

Composition of the Earth's Crust			
Element	Symbol	Percent by mass	Percent by volume
Oxygen	O	46.40	94.04
Silicon	Si	28.15	0.88
Aluminum	Al	8.23	0.48
Iron	Fe	5.63	0.49
Calcium	Ca	4.15	1.18
Sodium	Na	2.36	1.11
Magnesium	Mg	2.33	0.33
Potassium	K	2.09	1.42
All others	—	0.66	0.07

Figure 11-10. Percentages of the chief elements in the earth's crust by mass and by volume. "Volume" in this case means the amount of space occupied by the atoms of each element in the solid substances of the crust. Also see *Average Chemical Composition of Earth's Crust, Hydrosphere, and Troposphere* in the *Earth Science Reference Tables*

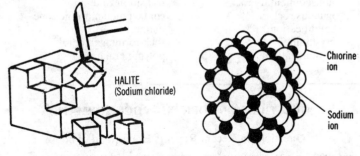

Figure 11-11. Cleavage of the mineral halite. When a piece of halite is broken, it tends to break into roughly cubic pieces. This is the result of a cubic arrangement of the constituent units (sodium and chlorine ions) in the halite crystal.

MINERAL STRUCTURE.   In the group of minerals called silicates, the constituent unit is in the form of a *tetrahedron* composed of silicon and oxygen (see Figure 11-12). This **silicon-oxygen tetrahedron** can combine with itself and with other elements to form different structures (such as chains and sheets). These different structural arrangements account for differences in the physical properties of silicate minerals.

Figure 11-12. The silicon-oxygen tetrahedron. A tetrahedron is a geometric figure having four faces, all of which are equilateral triangles of the same size. In silicates, there are oxygen atoms at the four corners of a tetrahedron, with a silicon atom at the center.

VOCABULARY

texture	intrusion
sedimentary rock	intrusive (plutonic) igneous rock
cementation	extrusion
compression/compaction	extrusive (volcanic) igneous rock
precipitation	metamorphic rock
monomineralic	regional metamorphism
crystal	recrystallization
chemical sedimentary rock	foliation
organic	distorted structure
beds/strata	rock cycle
nonsedimentary rock	transition zone
igneous rock	contact metamorphic zone
solidification	mineral
crystalline	rock-forming mineral
crystallization	polymineralic

QUESTIONS ON TOPIC XI—ROCK FORMATION

Questions in Recent Regents Exams (end of book)

Questions from Earlier Regents Exams

1. What do most igneous, sedimentary, and metamorphic rocks have in common?   (1) They are formed from molten material. (2) They are produced by heat and pressure.   (3) They are composed of minerals.   (4) They exhibit crystals, banding, and distinct layers.

2. In which rock type are fossils usually found?   (1) igneous (2) volcanic   (3) sedimentary   (4) metamorphic

3. Which rock was formed by the compaction and cementation of particles 0.07 centimeter in diameter?   (1) limestone   (2) sandstone (3) shale   (4) basalt

4. Which sedimentary rocks are formed from organic matter? (1) rock salt and shale   (2) bituminous coal and limestone   (3) chert and rock salt   (4) sandstone and conglomerate

5. According to the *Earth Science Reference Tables*, limestone, gypsum, and rock salt are rocks formed by the process of   (1) melting and solidification   (2) evaporation and precipitation   (3) erosion and deposition   (4) weathering and metamorphism

6. Which would most likely occur during the formation of igneous rock?   (1) compression and cementation of sediments   (2) recrystallization of unmelted material   (3) solidification of molten materials (4) evaporation and precipitation of sediments

7. Which graph best shows the relationship between the size of the crystals in an igneous rock and the length of time it has taken the rock to solidify?

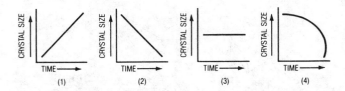

8. According to the *Earth Science Reference Tables*, generally, as the percentage of felsic minerals in a rock increases, the rock's color will become (1) darker and its density will decrease   (2) lighter and its density will increase   (3) darker and its density will increase (4) lighter and its density will decrease

9. An igneous rock which has crystallized deep below the earth's surface has the following approximate composition: 70% pyroxene, 15% plagioclase, and 15% olivine. According to the *Earth Science Reference Tables*, what is the name of this igneous rock?   (1) granite   (2) rhyolite (3) gabbro   (4) basalt

10. Which diagram below shows an area in which fine-grained igneous rock are most likely to be found?

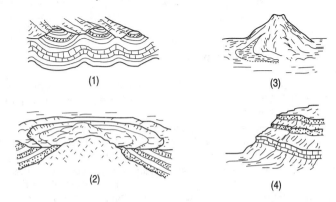

(1)

(3)

(2)

(4)

11. According to the *Earth Science Reference Tables*, metamorphic rocks form as the direct result of   (1) precipitation from evaporating water   (2) melting and solidification in magma   (3) erosion and deposition of soil particles   (4) heat and pressure causing changes in existing rock.

12. The diagram below represents a cross section of a coarse-grained nonsedimentary rock. According to the *Earth Science Reference Tables*, this rock is most likely to be   (1) basalt   (2) rhyolite   (3) gabbro   (4) granite

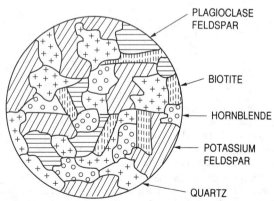

PLAGIOCLASE FELDSPAR

BIOTITE

HORNBLENDE

POTASSIUM FELDSPAR

QUARTZ

13. Which rock is composed of materials that show the greatest variety of rock origins?   (1) a limestone composed of coral fragments cemented together by calcium carbonate   (2) a conglomerate composed of pebbles of granite, siltstone, and basalt   (3) a very fine-grained basalt with sharp edges   (4) a sandstone composed of rounded grains of quartz

14. What is the main difference between metamorphic rocks and most other rocks? (1) Many metamorphic rocks contain only one mineral. (2) Many metamorphic rocks have an organic composition. (3) Many metamorphic rocks exhibit banding and distortion of structure. (4) Many metamorphic rocks contain a high amount of oxygen-silicon tetrahedra.

15. The green sand found on some Hawaiian Island shorelines most probably consists primarily of (1) quartz (2) olivine (3) plagioclase feldspar (4) potassium feldspar

16. Large rock salt deposits in the Syracuse area indicate that the area once had (1) large forests (2) a range of volcanic mountains (3) many terrestrial animals (4) a warm, shallow sea

17. According to the Rock Cycle diagram in the *Earth Science Reference Tables*, which type(s) of rock can be the source of deposited sediments? (1) igneous and metamorphic rocks, only (2) metamorphic and sedimentary rocks, only (3) sedimentary rocks, only (4) igneous, metamorphic, and sedimentary rocks

18. The data table shows the composition of six common rock-forming minerals.

Mineral	Composition
Mica	$KAl_3Si_3O_{10}$
Olivine	$(FeMg)_2SiO_4$
Orthoclase	$KAlSi_3O_8$
Plagioclase	$NaAlSi_3O_8$
Pyroxene	$CaMgSi_2O_6$
Quartz	SiO_2

The data table provides evidence that (1) the same elements are found in all minerals (2) a few elements are found in many minerals (3) all elements are found in only a few minerals (4) all elements are found in all minerals

19. The cubic shape of a crystal is most likely the result of that crystal's (1) hardness (2) density distribution (3) internal arrangement of atoms (4) intensity of radioactive decay

20. Minerals are identified on the basis of (1) the method by which they were formed (2) the type of rock in which they are found (3) the size of their crystals (4) their physical and chemical properties

21. Which property is most useful in mineral identification? (1) hardness (2) color (3) size (4) texture

22. The hardness of a mineral such as quartz is due to (1) the internal arrangement of its atoms (2) large amounts of impurities (3) its characteristic luster and color (4) its formation in certain rock types

23. Which object is the best model of the shape of a silicon-oxygen structural unit?

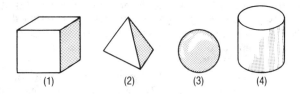

(1)   (2)   (3)   (4)

Additional Questions

To answer questions 1 through 7 refer to the diagrams below, which represent six rock samples. The shadings indicate different minerals, sediments, or fossils.

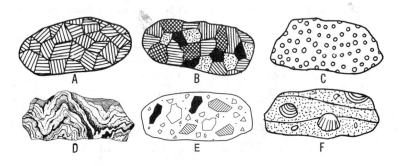

1. Which rock would be considered monomineralic?   (1) A   (2) B   (3) D   (4) E

2. Which rock formed partly as the result of biologic processes?   (1) B   (2) C   (3) D   (4) F

3. Which sample is nonsedimentary?   (1) B   (2) C   (3) E   (4) F

4. Which sample has distorted structure and foliation?   (1) A   (2) B   (3) C   (4) D

5. Which sample is most likely breccia?   (1) B   (2) C   (3) E   (4) F

6. If sample E were subjected to high heat and pressure, without melting, it would most likely form   (1) slate   (2) schist   (3) metaconglomerate   (4) marble

7. Rock D is most likely   (1) gneiss   (2) siltstone   (3) slate   (4) pumice

TOPIC XII — The Dynamic Crust (Lithosphere) and the Earth's Interior

There is much evidence indicating that the crust is constantly undergoing change, and that all parts of it move to new locations, thereby changing the surface of the earth. In some cases we know the crust moves because its movement has been observed directly during earthquakes or the sudden birth of volcanoes. Evidence of past crustal movements must be indirect. Much of that indirect evidence is based on the concept of **original horizontality,** which assumes (on the basis of geologic evidence) that sedimentary rocks and some extrusive igneous rocks form in generally horizontal layers, or **strata.** Therefore, most strata found in other than horizontal positions are believed to have been deformed by crustal movement. Note: In this topic whatever is stated about the crust also applies to the lithosphere, except when the terms "granitic crust" or "basaltic crust" are used.

EVIDENCE OF MINOR CRUSTAL CHANGES

DEFORMED STRATA.   Rock strata that no longer show their original horizontality are called *deformed strata*. Some of the types of deformed strata are **folded strata, tilted strata,** and **faulted strata** (see Figure 12-1). All such deformed strata are considered evidence of past crustal movements. A **fault** is a crack in a mass of rock along which there has been *displacement*, or movement, of the rock layers. Movement along a fault, or faulting, results in a shaking of the earth called an earthquake.

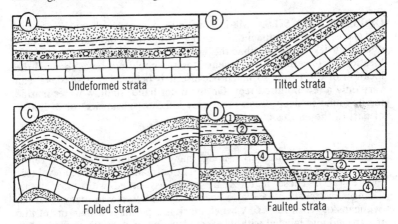

Figure 12-1. Three types of deformed horizontal strata. The numbers in Diagram D mark strata that were originally continuous.

MOVEMENTS DURING EARTHQUAKES. It is sometimes possible to observe the actual movement of the crust or immediate effects of that movement during an earthquake. This occurs when the movement along a fault results in obvious changes in the positions of surface features, sometimes forming cliffs (see Figure 12-2).

Figure 12-2. Displacement of surface features as the result of movement of strata along a fault. The portions of the power line, road, orchard, and fields in the foreground have moved down and to the right relative to their continuations at the top of the cliff. Displacements during a single fault movement seldom exceed a few meters. However, successive movements over millions of years may produce accumulated displacements of 1,000 meters or more.

DISPLACED FOSSILS. Marine (meaning ocean) fossils, such as corals and clams, found in sedimentary rock high (sometimes thousands of meters) above sea level indicate that the land has been uplifted to its present location. A lowering of the level of the sea could not have changed the location of the fossils so much because it is believed that sea level can vary only a few hundred feet. On the other hand, shallow water marine fossils found in deep ocean areas may indicate **subsidence,** or a sinking of part of the earth's crust.

VERTICAL MOVEMENTS. Along many shorelines there are raised beaches and other coastline features that normally occur at sea level (see Figure 12-3). These features are often associated with tilted rocks. Changes in elevation have also been observed directly in many locations by means of *bench marks.* A **bench mark** is a permanent marker set into the ground and labeled with its exact elevation at the time of placement. Later measurements often show changes in elevation. All these changes indicate that vertical movements of the earth's crust have taken place. Vertical movements often affect large portions of the earth's crust.

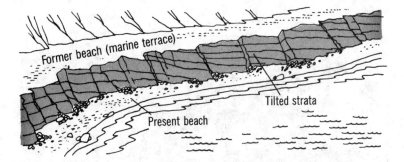

Figure 12-3. Evidence of vertical movement of the crust along a shoreline. The flat marine terrace can be recognized as having once been a beach. The beach was raised by crustal movement. When movement stopped, the ocean once again began to cut a level beach into the new shore.

EVIDENCE OF MAJOR CRUSTAL CHANGES

GEOSYNCLINES AND MOUNTAINS. Large thicknesses of sedimentary rock have been found in parts of various mountain systems. These rocks contain fossils and other evidence which indicate that they formed in shallow marine (ocean) water. For this reason, many geologists believe that the continental mountain systems formed from *geosynclines*. A **geosyncline** is a large shallow ocean basin, located near the margin of a continent, that subsides, or sinks, as sediments are added to it. This subsidence would explain how the great thicknesses of sediments needed to form mountains can be deposited in a relatively shallow geosyncline without filling it. In this way, the geosyncline can remain a shallow-water area for millions of years. Later, the sedimentary rock strata of the geosyncline are believed to be deformed and uplifted to form mountains and enlarge the continent. In terms of modern plate tectonics, an ocean-continent margin not at a plate boundary is a basin similar to a geosyncline called a **passive margin basin**. Figure 12-4 on page 164 indicates some possible geosynclines, that is, large shallow basins where sedimentation and possibly subsidence are presently occurring.

ISOSTASY AND GEOSYNCLINES. The principle of **isostasy** has been used to explain part, but not all, of the sinking of geosynclines. The principle states that the lithosphere of the earth is in a state of equilibrium, and that any change in mass of one part of the lithosphere will be offset by a change in mass of another part to maintain the equilibrium. Think of the lithosphere as floating on the denser rock of the mantle as an ice cube floats in water. If erosion causes the continental crust to lose sediments to a geosyncline, the continental crust will slowly rise and the geosyncline will slowly sink. This occurs because the continent has lost mass and therefore floats higher, while the geosyncline has gained mass and therefore floats lower (see Figure 12-5 on page 165). Isostasy is probably due partly to the partly liquid, plasticlike properties of the upper mantle layer called the **asthenosphere**.

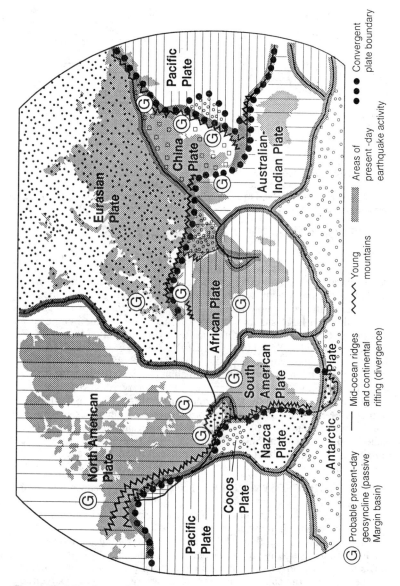

Figure 12-4. Regions of the earth's major plates and zones of current crustal activity. Note that the borders of the plates are the locations of most current crustal activity, which includes volcanic eruptions and earthquakes. Most of the world's active volcanoes are located where mid-ocean ridges or young mountains are indicated on the map.

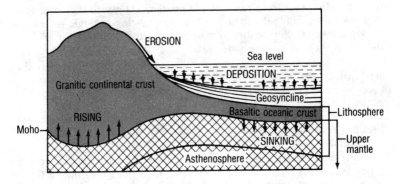

Figure 12-5. The theory of isostasy. As the geosyncline (passive margin basin) becomes heavier because of the accumulation of sediments, the oceanic crust sinks lower into the mantle. At the same time, the continental crust becomes lighter because of the erosion of material, and it rises higher. Note that where elevations are highest the crust is thickest.

ZONES OF CRUSTAL ACTIVITY AND OROGENY. Zones of frequent crustal activity can be located on the earth's surface. Earthquakes, volcanoes, ocean trenches, mid-ocean ridges (mountain ranges in the ocean), and young continental mountains occur in these zones, or belts, as shown in Figure 12-4. The fact that these features are found together indicates that they are closely related.

At some times large amounts of crustal activity result in mountain building. The process of mountain building is called **orogeny.**

CONTINENTAL DRIFT. In recent years much evidence has been presented to support the idea that the continents have been moving around on the earth's surface. This concept is called **continental drift.** Some of the evidence for continental drift is given below:

1. The outlines of the continents appear to fit together almost like a jigsaw puzzle. It is believed that at least twice in the geologic past all the continents were connected, and that for the last 230,000,000 years the continents have mainly moved apart. (See Figure 12-6 on page 166 and *Geologic History of New York State at a Glance* in the *Earth Science Reference Tables.*)

2. The present orientations of rock crystals with magnetic properties also give evidence of crustal shifts. When an igneous rock crystallizes from the molten state, any of its minerals that are magnetic will align themselves with the earth's magnetic field. By studying the alignment of these crystals, we can tell the relative direction of the earth's magnetic poles at the time the rock formed. When rocks of different geological times are studied in this way, it is found that the earth's magnetic poles *seem* to have wandered around the earth. However, it is more likely that

the magnetic poles have remained approximately where they are, and that it was the gradual shifting of the crust during geologic time that caused crystals to change direction, the change being greater the greater the age of the rock. (See Figure 12-6.)

3. At places where the continents may have fitted together, the similarity of the minerals, fossils, and rock types indicates that these rocks of the different continents are similar in age and origin. The northern Appalachian Uplands (see Figure 14-2, page 205) contain many rock types, structures, and fossils similar to those in parts of western Europe. This suggests that the continents were together when the rocks were formed.

4. Today the continents are separated and their respective life forms are greatly different. However, fossil evidence shows that at times in the past many land (terrestrial) plants and animals were the same throughout the world. Similar dinosaurs lived on most land areas during the Triassic Period, but not during the Cretaceous Period (see the *Geologic History of New York State at a Glance* in the *Earth Science Reference Tables*). Such a wide distribution of the same plants and animals probably could not have occurred unless the continents were connected.

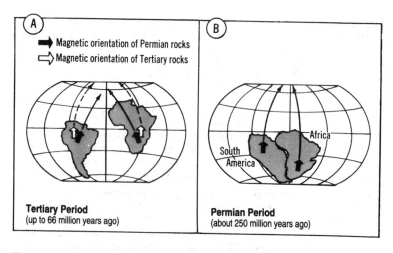

Figure 12-6. Evidence of continental drift. The present continents have shapes that suggest that they once were parts of larger landmasses, which have drifted apart by crustal movement. Magnetic orientation of rock crystals supports this hypothesis. Diagram A shows the magnetic orientation of rocks of different ages in South America and Africa. Crystals in rocks of relatively recent formations are magnetically oriented toward the earth's magnetic poles. Older rocks are oriented in different directions. If, however, it is assumed that South America and Africa were parts of a single landmass when these older rocks were forming, as shown in Diagram B, then the magnetic orientation of these older formations is found to agree reasonably well.

OCEAN-FLOOR SPREADING. Much evidence indicates that the sea floor has spread out, or **rifted**, from the **mid-ocean ridge** systems (mountain ranges in the ocean). This concept is called **ocean-floor spreading** or sea-floor spreading. The location of mid-ocean ridges is shown in Figure 12-4. Here is some of the evidence for ocean-floor spreading:

  1. Molten rock rises and crystallizes into basaltic igneous rock in the mid-ocean ridges, as shown by high heat flows and observed volcanic activity. Samples of this basaltic rock taken from the ocean floor show that the age of the rock increases with distance from the center of the mid-ocean ridge. Heat flow also decreases with distance from the mid-ocean ridges in oceanic crust. This indicates that the basaltic rock formed at the ridges and spread outward (see Figure 12-7A, page 168).

  2. The spreading of the oceanic crust is also shown by the magnetism in the basaltic rocks. The earth's magnetic poles flip-flop in polarity (north changes to south or south to north) in periods of thousands of years in a process called **reversal of earth's magnetic polarity.** The earth's magnetic fields have reversed polarity thousands of times since its origin. When the basaltic rock crystallizes at the mid-ocean ridges, its magnetic minerals are aligned so that they record the particular polarity of the earth at the time of crystallization. It has been found that there are corresponding parallel strips of basaltic rock on either side of the mid-ocean ridges. Some strips show normal polarity (poles as they are today) and others show reverse polarity. This evidence suggests that the corresponding strips were formed at similar times and that ocean-floor spreading has separated them.

PLATE TECTONICS AND CRUSTAL CHANGE. Since the total mass and volume of the earth are fixed, the creation of new areas of oceanic crust by ocean-floor spreading requires that the crust be destroyed elsewhere at the same rate. As part of a major revolution in scientific thought about the earth's structure, geologists are attempting to explain crustal changes in terms of the creation, motion, and destruction of large sections of the lithosphere called **plates.** This revolutionary idea is called the **plate tectonic theory.** It includes the concepts of continental drift and ocean-floor spreading. (*Tectonics* is a term that applies to changes in the earth's crust and the forces that cause them.)

  According to the plate tectonic theory, the earth's lithosphere is divided into an irregular pattern of solid, moving plates that are created at one edge and destroyed at the other edge (see Figures 12-4 and 12-7). The tectonic plates and their associated continents are thought to be moving at rates of approximately a few centimeters a year. The plates move on the plasticlike layer of the upper mantle called the asthenosphere.

  Earthquakes, volcanoes, ocean trenches, and young continental mountains occur together, as shown in Figure 12-4. It is believed that these zones of frequent crustal activity mark the boundaries of plates. The features and events of these zones are assumed to be the result of plate movements.

Figure 12-7 shows the three major types of plate boundaries. One of these boundary types is where two plates are spreading apart at a mid-ocean ridge. This is called a **divergent plate boundary.** The spreading is due to plate growth from volcanic activity. Another type of plate boundary is where plates collide. This is called a **convergent plate boundary.** The colliding edges of the plates become distorted, and young continental mountains may form between them. An existing passive margin basin along a continental margin near the boundary of colliding plates would then be uplifted and changed into folded mountains. Often in plate collisions the edge of one plate is pushed down under the other. The subsiding plate plunges or subducts down into the hot mantle, where it is destroyed. The **subduction** and resultant melting of the plate forms liquid rock, which rises and produces the many volcanic and intrusive features associated with plate collisions. *Oceanic trenches* and *volcanic island arcs* often occur in areas of subduction. The third type of plate boundary occurs where the plates slide by each other in a sideways motion. This is called a **lateral (or transform) fault plate boundary.** The San Andreas fault in California is an example of a lateral fault plate boundary.

PLATE TECTONICS AND MANTLE CONVECTION. In order for tectonic plates to move, it is assumed that the mantle under the lithosphere and asthenosphere must behave like a fluid. Some geologists believe that there are convection currents within the mantle that cause plate movement. A proposed source of energy for these currents is the decay of radioactive materials within the earth. One form of evidence for these mantle convection currents is that there are variations in heat flow from the earth. The heat flow at the mid-ocean ridge, where it is believed the convection currents are rising, is higher than average, and heat flow decreases as distance from the mid-ocean ridges increases.

EARTHQUAKES

An **earthquake** is natural, rapid shaking of the lithosphere caused by displacement of rock, such as along faults. The energy of an earthquake is transferred away from its point of origin, which is called the **focus** (plural: *foci*). The energy is transmitted as **seismic** or earthquake **waves,** which are studied by scientists using an instrument called a **seismograph.**

Figure 12-7. Types of plate boundaries. (A) Diverging (rifting) plate boundary with a mid-ocean ridge, shallow-depth earthquakes, igneous intrusions, extrusions of lava flows, and volcanoes. Plate 1 is moving west and Plate 2 is moving east. **(B)** Converging plate boundary with an oceanic trench, volcanic island arcs, igneous intrusions, young mountains, and subduction of oceanic plate. Earthquake foci exist at various depths, indicating subduction. Plate 1 is moving east and Plate 2 is moving west. **(C)** Lateral (transform) fault plate boundary where there is sideways movement of the two plates. This boundary is associated with many shallow-focus earthquakes, and no igneous activity.

A. Diverging Plate Boundary

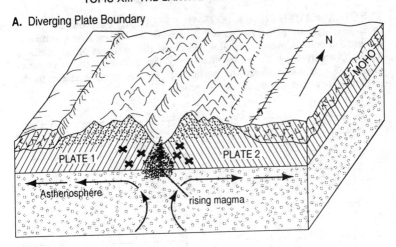

B. Converging Plate Boundary

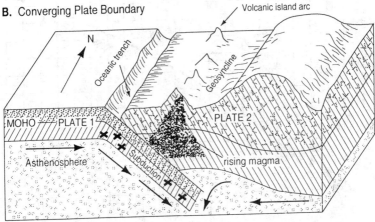

C. Lateral Transform Plate Boundary

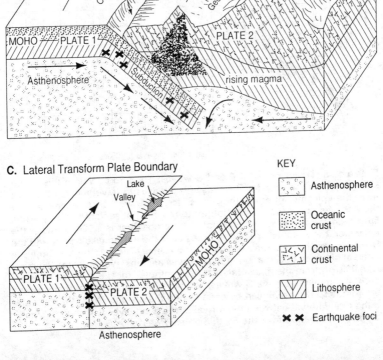

KEY

░	Asthenosphere
⬚	Oceanic crust
⩗	Continental crust
⑊	Lithosphere
✖ ✖	Earthquake foci

TYPES OF EARTHQUAKE (SEISMIC) WAVES. Earthquakes create **compressional,** or **primary,** waves **(P-waves),** and **shear,** or **secondary,** waves **(S-waves).** Compressional waves cause the particles of the material through which they travel to vibrate in the direction the waves are moving. Shear waves cause the particles to vibrate at right angles to the direction in which the waves are moving.

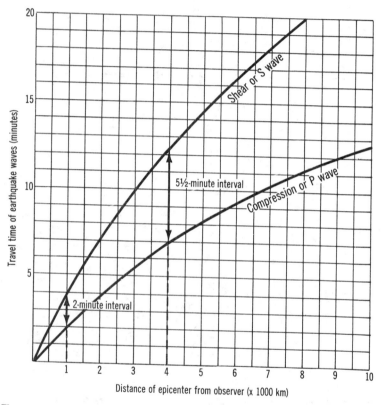

Figure 12-8. Travel times of P and S waves. The two curves in this graph show the time required for P waves and S waves to travel a given distance from the epicenter of an earthquake. Since the S waves travel slower than the P waves, the S waves take longer to reach an observer, and they arrive later than the P waves. For example, if the observer is 1,000 km from the epicenter, the P waves arrive 2 minutes after the earthquake occurs and the S waves arrive 4 minutes after the occurrence. There is thus a 2-minute interval between observation of the P waves and observation of the S waves. At 4,000 km from the epicenter, it takes 7 minutes for the P waves to arrive and 12½ minutes for the S waves; the time interval between them is thus 5½ minutes. The graph can be used to find the distance from the epicenter if the time interval between the arrival of the two waves is known. A similar diagram can be found in the *Earth Science Reference Tables.*

PROPERTIES OF EARTHQUAKE WAVES

1. In any one medium, *compressional waves travel faster than shear waves*. Thus when an earthquake occurs, compressional waves will reach a seismograph before shear waves (see Figure 12-8).

2. The velocity of seismic waves in the earth depends upon the physical properties of the material they are passing through. Generally, the more dense the material, the greater the velocity of the waves. As they pass from a material of one density to a material of another density, seismic waves are bent or refracted.

3. Within the same material, an increase in pressure increases the velocity of seismic waves.

4. Compressional waves will pass through solids, liquids, and gases, while shear waves will pass through solids only.

LOCATION OF AN EPICENTER.   The **epicenter** of an earthquake is the place on the earth's surface directly above or closest to the point of origin, or focus, of the earthquake. Epicenters are located by using the velocity differences between compressional (P-waves) and shear (S-waves). P-waves move faster than S-waves; therefore, the farther an observer is from an epicenter, the larger the *time interval* between the arrival of the P-waves and S-waves. The distance to the epicenter is determined by comparing the interval with the graph data.

To find the position of the epicenter, at least three seismograph locations must be used, and epicenter distances must be calculated for each. For each of the three locations, the epicenter distance is then used as a radius and circles are drawn on a globe or map, as shown for locations A, B, and C in Figure 12-9. The place where all three circles intersect is the epicenter of the earthquake.

FINDING THE ORIGIN TIME OF EARTHQUAKES. The time at which an earthquake originates can be determined from the epicenter distance and seismic-wave travel time. The farther an observer is from the epicenter, the longer it takes the seismic waves to travel to the observation point. For example, suppose the observer is 4,000 kilometers from the epicenter. Figure 12-8 shows that it took the P-wave seven minutes to arrive; thus the earthquake occurred seven minutes earlier than the time at which the P-waves were observed on the seismograph. As another example, if the S-wave first arrived at a station at 10 hr:12 min:30 sec G.M.T. (Greenwich Mean Time) and the seismograph station is 5,500 kilometers away from the epicenter, when did the earthquake occur?

A MODEL OF THE EARTH'S INTERIOR

ZONES OF THE EARTH.   Analysis of seismic waves indicates that the earth is composed of a number of zones. The **crust** and the upper part of the **mantle** (the rigid mantle) make up the lithosphere. The boundary between the crust and the rigid mantle is called the Moho. The rocks of the lithosphere become more dense below this boundary. The asthenosphere is the

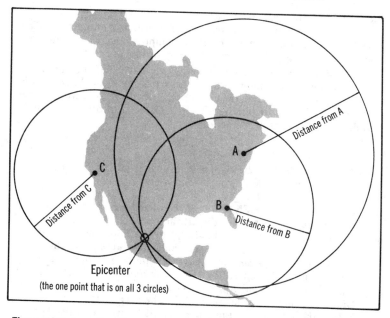

Figure 12-9. Locating the epicenter of an earthquake. Seismograph observations at three widely spaced locations are needed. For the time interval between the P and S waves, the distance to the epicenter can be determined for each station. A circle with this radius is then drawn around the station on a map or globe. The epicenter must lie somewhere on this circle. The same is done for all three stations. The one point that lies on all three circles must be the epicenter.

partly molten, plasticlike part of the mantle. Below it are the stiffer part of the mantle, the **outer core,** and the **inner core.**

Figure 12-10 shows the approximate dimensions and locations of these zones and how they relate to the velocity and type of earthquake waves. *The outer core is believed to be liquid because the S-waves do not penetrate it and because of the sharp decrease in P-wave velocity.* Because S-waves do not pass through the outer core, there is an area on the side of the earth away from the focus of an earthquake where S-waves are not recorded. This area is known as the earthquake's shadow zone.

As depth into the earth increases, temperature, pressure, and density generally increase. Be sure to study the diagram *Inferred Properties of the Earth's Interior* in the *Earth Science Reference Tables.*

EARTH'S CRUST. The earth's crust, or upper lithosphere, is divided into two major divisions: the **continental crust** and the **oceanic crust.** The continental crust is usually much thicker than the oceanic crust (see Figure 12-5). The crust is thickest where it is highest—in mountain regions.

The two crusts are also distinguished by differences in composition and density. The continental crust is made mostly of granitic rocks, rocks with a composition similar to granite. The oceanic crust is composed mostly of basaltic rocks, rocks similar in composition to basalt. The granitic continental crust is less dense than the basaltic oceanic crust, as shown on the *Scheme for Igneous Rock Identification* in the *Earth Science Reference Tables.*

COMPOSITION OF THE EARTH'S INTERIOR.   The iron and nickel composition of metallic meteorites suggests that the composition of the earth's interior is mostly iron and nickel. A combination of iron and nickel at the temperature and pressure believed to be in the earth's core can account for some of the observed properties of seismic waves in the core. The high-density iron-nickel composition of the core and the low-density oxygen-silicon composition of the crust indicate that the mantle must have a composition different from the crust and the core.

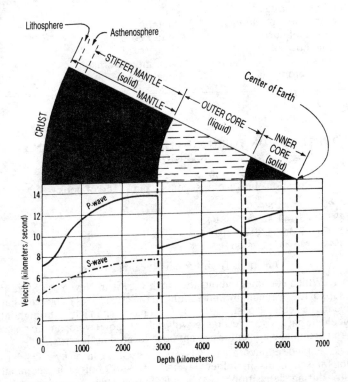

Figure 12-10. Zones of the earth's structure as deduced from characteristics of P and S waves. (Also see *Inferred Properties of the Earth's Interior* in the *Earth Science Reference Tables.*)

VOCABULARY

original horizontality
strata
folded strata
tilted strata
faulted strata
fault
subsidence
bench mark
geosyncline
passive margin basin
isostasy
asthenosphere
continental drift
orogeny
mid-ocean ridge
ocean-floor spreading
reversal of earth's
  magnetic polarity
plates
plate tectonic theory

divergent plate boundary
convergent plate boundary
subduction
lateral   (or transform) fault
  plate boundary
earthquake
focus (of earthquake)
seismic waves
seismograph
compressional wave
  (primary or P-wave)
shear wave (secondary
  or S-wave)
epicenter
crust
mantle
outer core
inner core
continental crust
oceanic crust

QUESTIONS ON TOPIC XII—THE EARTH'S CRUST AND INTERIOR

Questions in Recent Regents Exams (end of book)

Questions from Earlier Regents Exams

1. Which statement about the earth's crust in New York State is best supported by the many faults found in the crust?   (1) The crust has moved in the geologic past.   (2) The crust has been inactive throughout the geologic past.   (3) New faults will probably not develop in the crust. (4) An earthquake epicenter has not been located in the crust.

2. Folded sedimentary rock layers are usually caused by   (1) deposition of sediments in folded layers.   (2) differences in sediment density during deposition   (3) a rise in sea level after deposition (4) crustal movement occurring after deposition

3. The diagram below represents a cross section of a portion of the earth's crust. What do these tilted rock layers suggest?

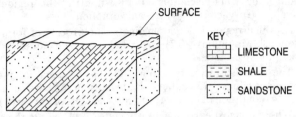

(1) This area remained fairly stable since the sediments were deposited. (2) The sediments were deposited at steep angles and then became rock.   (3) Metamorphism followed the deposition of the sediments. (4) Crustal movement occurred some time after the sediments were deposited.

4. The diagram below represents a vertical cross section of sedimentary rock layers which have not been overturned. Which principle best supports the conclusion that these layers have undergone extensive movement since deposition?

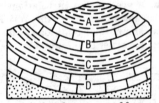

(1) Sediments are deposited with the youngest layers on top.   (2) Sediments are deposited in horizontal layers.   (3) Rock layers are older than igneous intrusions.   (4) Sediments containing the remains of marine fossils are deposited above sea level.

5. The landscape shown in the diagram below is an area of frequent earthquakes.

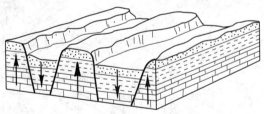

This landscape provides evidence for   (1) converging convection cells within the rocks of the mantle   (2) density differences in the rocks of the mantle   (3) movement and displacement of the rocks of the crust   (4) differential erosion of hard and soft rocks of the crust

6. Which is the most logical conclusion to be drawn from marine fossils found on a mountaintop?   (1) marine animals once lived on land (2) sediments once under water were uplifted   (3) the fossils were transported there   (4) the mountain is composed of igneous rock

7. A line of former beaches along a coast, all 50 meters above sea level, is evidence of   (1) present erosion   (2) the present melting of polar ice caps   (3) land uplift   (4) a decrease in the deposition of marine fossils

8. The thick sedimentary rocks of central and western New York State, which were formed from shallow water deposits, were most probably produced by   (1) glaciation   (2) the uplift of this region   (3) deposition in a geosyncline   (4) volcanic eruption

9. In which region is a geosyncline in the process of formation? (1) Adirondack Mountains   (2) Great Lakes   (3) Colorado Plateau (4) Mississippi River delta

10. The term isostasy refers to a   (1) line of equal air pressure (2) series of anticlines   (3) deflection of winds on the earth caused by rotation   (4) condition of balance between segments of the earth's crust

11. Where do most earthquakes originate?   (1) within the earth's outer core   (2) along specific belts within the crust   (3) randomly across the entire earth's surface   (4) evenly spaced along the Moho interface

Base your answers to questions 12 through 16 on the diagram below. The diagram is a model which represents one possible interpretation of the movements of the earth's rock surfaces according to the theory of plate tectonics (continental drift and sea floor spreading). According to this interpretation, the earth's lithosphere consists of several large "plates" which are moving in relationship to one another. The arrows in the diagram show some of this relative motion of the "plates." The diagram also shows the age of formation of the igneous rocks that make up the oceanic crust of the northern section of the Pacific Plate.

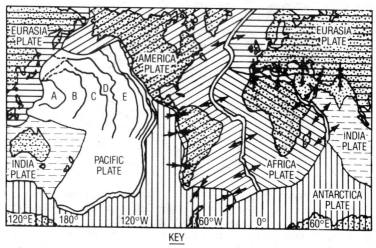

KEY

→|← Movement of plates toward each other

←| |→ Movement of plates away from each other

AGE OF ROCKS OF PACIFIC PLATE

A Jurassic   B Early Cretaceous   C Middle Cretaceous   D Late Cretaceous   E Eocene

12. Which statement is best supported by the relative movement shown by the arrows in the diagram?   (1) North America and South America are moving toward each other.   (2) The India Plate is moving away from the Eurasia Plate.   (3) The Africa Plate and Eurasia Plate are moving away from the America Plate.   (4) The Antarctica Plate is moving away from the America Plate.

13. The boundaries between all of these "plates" are best described as the sites of   (1) frequent crustal activity   (2) deep ocean depths   (3) continental boundaries   (4) magnetic field reversals

14. Which geologic structure is represented by the double line separating the America Plate from the Africa and Eurasia Plates?   (1) thick continental crust   (2) thick layers of sediment   (3) mid-ocean ridge   (4) granitic igneous rock

15. Which provides the best explanation of the mechanism that causes these "plates" to move across the earth's surface?   (1) convection currents in the mantle   (2) faulting of the lithosphere   (3) the spin of the earth on its axis   (4) prevailing wind belts of the troposphere

16. The age of formation of the igneous rocks A, B, C, D, and E which make up the oceanic crust of the northern half of the Pacific Plate suggests that this section of the Pacific Plate is generally moving in which direction?   (1) from north to south   (2) from south to north   (3) from west to east (4) from east to west

17. Recent volcanic activity in different parts of the world supports the inference that volcanoes are located mainly in   (1) the centers of landscape regions   (2) the central regions of the continents   (3) zones of crustal activity   (4) zones in late stages of erosion

18. Which evidence does *not* support the theory that Africa and South America were once part of the same large continent?   (1) correlation of rocks on opposite sides of the Atlantic Ocean   (2) correlation of fossils on opposite sides of the Atlantic Ocean   (3) correlation of coastlines on opposite sides of the Atlantic Ocean   (4) correlation of living animals on opposite sides of the Atlantic Ocean

19. Which statement best supports the theory of continental drift?   (1) Basaltic rock is found to be progressively younger at increasing distances from a mid-ocean ridge.   (2) Marine fossils are often found in deep-well drill cores.   (3) The present continents appear to fit together as pieces of a larger landmass.   (4) Areas of shallow-water seas tend to accumulate sediment, which gradually sinks

20. According to the *Earth Science Reference Tables*, during which geologic time period were the continents of North America, South America, and Africa closest together?   (1) Tertiary   (2) Cretaceous   (3) Triassic   (4) Ordovician

21. According to the concept of continental drift, the distance between two continents on opposite sides of a mid-oceanic ridge will generally (1) decrease   (2) increase   (3) remain the same

22. Two samples of ocean floor basaltic bedrock are found at equal distances from, and on opposite sides of, a mid-ocean ridge. The best evidence that both were formed at the ridge during the same time period would be that both samples also   (1) have the same density   (2) contain different crystal sizes   (3) are located at different depths below sea level   (4) have the same magnetic field orientation

23. Which of the cross-sectional diagrams below best represents a model for the movement of rock material below the crust along the mid-Atlantic ridge?

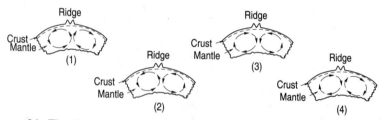

24. The drawing below represents the ocean floor between North America and Africa.

Which graph best represents the age of the bedrock in the ocean floor along line *AB*?

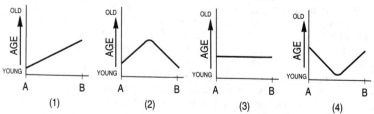

25. Which observation provides the strongest evidence for the inference that convection cells exist within the earth's mantle?   (1) Sea level has varied in the past.   (2) Marine fossils are found at elevations high above sea

level.   (3) Displaced rock strata are usually accompanied by earthquakes and volcanoes.   (4) Heat-flow readings vary at different locations in the earth's crust.

26. Placing a seismograph on the moon enables us to determine if the moon has   (1) water   (2) an atmosphere   (3) radioactive rock   (4) crustal movements

27. The place on the earth's surface directly above the point at which an earthquake originates is the   (1) epicenter   (2) focus   (3) Moho   (4) zenith

28. The immediate result of a sudden slippage of rocks within the earth's crust will be   (1) isostasy   (2) an earthquake   (3) erosion   (4) the formation of convection currents

29. The action of particles as compression waves (P-waves) pass is most nearly like   (1) a line of parked cars being hit from behind by a moving car   (2) the bobbing of corks on the surface of water   (3) the twisting of a rubber hose   (4) the spinning of a top

30. A characteristic of compressional waves and shear waves (S-waves) is that they both   (1) travel at the same speed   (2) travel faster through more dense solid materials   (3) travel through liquid and solid materials   (4) cause rock particles to vibrate in the same direction

The diagram illustrates how the epicenter of an earthquake is located by observatories in Pasadena, California; Chicago, Illinois; and Washington, D.C. Base your answers to questions 31 through 33 on the diagram.

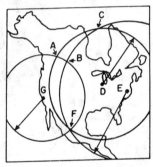

31. The epicenter of the earthquake is located nearest letter   (1) A   (2) B   (3) C   (4) F

32. The separation in time between the arrival of primary and secondary waves is   (1) greatest at Pasadena (G)   (2) greatest at Chicago (D)   (3) greatest at Washington, D.C. (E)   (4) the same at all stations

33. If the method illustrated by the diagram is used to locate the epicenter of an earthquake, it appears unlikely for an individual observatory, operating independently, to determine the   (1) direction to the epicenter   (2) distance to the epicenter   (3) distance to an epicenter located under the ocean   (4) interval between the initial and subsequent seismic waves

34. The time lapse between the arrival of the P-waves and S-waves on *one* seismograph recording can be used to determine the   (1) magnitude of the earthquake   (2) exact location of the focus   (3) exact location of the epicenter   (4) distance to the epicenter

Base your answers to questions 35 through 39 on the *Earth Science Reference Tables* and the diagram below. The diagram represents a cross section of the earth showing the paths of earthquake waves from a single earthquake source. Seismograph stations are located on the earth's surface at points *A* through *F*, and they are all located in the same time zone.

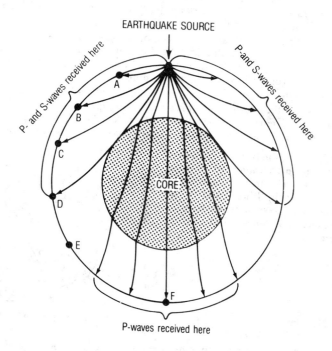

EARTHQUAKE SOURCE

P- and S-waves received here

P- and S-waves received here

CORE

P-waves received here

35. At which station is the distance in time between the arrival of P- and S-waves the greatest?   (1) *A*   (2) *B*   (3) *C*   (4) *D*

36. Station *E* did *not* receive any P-waves or S-waves from this earthquake because the P-waves and S-waves   (1) cancel each other out   (2) are bent, causing shadow zones   (3) are changed to sound energy   (4) are converted to heat energy

37. What explanation do scientists give for the reason that station *F* did *not* receive S-waves?   (1) The earth's inner core is so dense that S-waves cannot pass through.   (2) The earth's outer core is liquid, which does not allow S-waves to pass.   (3) S-waves do not have enough energy to pass completely through the earth.   (4) S-waves become absorbed by the earth's crust.

38. Seismograph station *D* is 7700 kilometers from the epicenter. If the P-wave arrived at this station at 2:15 P.M., at approximately what time did the earthquake occur?   (1) 1:56 P.M.   (2) 2:00 P.M.   (3) 2:04 P.M. (4) 2:08 P.M.

39. Seismograph station *B* recorded the arrival of P-waves at 2:10 P.M. and the arrival of S-waves at 2:15 P.M. Approximately how far is station *B* from the earthquake epicenter?   (1) 1400 km   (2) 2400 km (3) 3400 km (4) 4400 km

40. The diagrams below represent seismographic traces of three different disturbances A, B, and C recorded by the same seismograph.

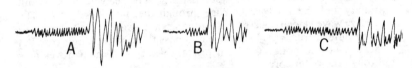

The traces indicate that the distance from the station to the epicenters of the three disturbances is   (1) least for disturbance A   (2) least for disturbance B   (3) least for disturbance C   (4) the same for all three disturbances

41. What total distance does an earthquake compression wave (P-wave) travel in its first 8 minutes?   (1) 2200 km   (2) 4850 km (3) 6500 km   (4) 11,200 km

42. A seismic station is 2000 kilometers from an earthquake epicenter. According to the *Earth Science Reference Tables*, how long does it take an S-wave to travel from the epicenter to the station?   (1) 7 minutes 20 seconds   (2) 5 minutes 10 seconds   (3) 3 minutes 20 seconds   (4) 4 minutes 10 seconds

43. In developing a model of the earth's deep interior, most of the evidence was derived from   (1) deep wells   (2) mining operations (3) observing other planets   (4) seismic data

44. Through which zones of the earth do compressional waves (P-waves) travel?   (1) crust and mantle, only   (2) mantle and outer core, only   (3) outer and inner core, only   (4) crust, mantle, outer core, and inner core

45. According to the *Earth Science Reference Tables*, in which group are the zones of the Earth's interior correctly arranged in order of increasing average density?   (1) crust, mantle, outer core, inner core   (2) crust, mantle, inner core, outer core   (3) inner core, outer core, mantle, crust   (4) outer crust, inner core, mantle, crust

46. According to the *Earth Science Reference Tables*, what is the relationship between density, temperature, and pressure inside the earth?   (1) As depth increases, density, temperature, and pressure decrease.   (2) As depth increases, density and temperature increase, but pressure decreases.   (3) As depth increases, density increases, but temperature and pressure decrease.   (4) As depth increases, density, temperature, and pressure increase.

47. The temperature of rock located 1000 kilometers below the Earth's surface is about   (1) 1000°C   (2) 2600°C   (3) 3300°C   (4) 4300°C

48. Which evidence best supports the inference that the earth's outer core possesses liquid characteristics?   (1) The velocities of both primary and shear waves increase through the outer core.   (2) The primary wave velocity decreases, while the shear wave velocity increases in the outer core.   (3) Primary waves pass through the outer core but shear waves do not.   (4) Both primary waves and shear waves pass through the outer core.

49. How does the composition of the oceanic crust compare with the composition of the continental crust?   (1) The oceanic crust is mainly limestone, while the continental crust is mainly sandstone.   (2) The oceanic crust is mainly limestone, while the continental crust is mainly granite.   (3) The oceanic crust is mainly basalt, while the continental crust is mainly sandstone.   (4) The oceanic crust is mainly basalt, while the continental crust is mainly granite.

50. As one travels from an ocean shore to the interior of a continent, the thickness of the earth's crust generally   (1) decreases   (2) increases   (3) remains the same

51. Which statement best describes the continental and oceanic crusts?   (1) The continental crust is thicker and less dense than the oceanic crust.   (2) The continental crust is thicker and more dense than the oceanic crust.   (3) The continental crust is thinner and less dense than the oceanic crust.   (4) The continental crust is thinner and more dense than the oceanic crust.

52. Which part of the earth is probably most responsible for its high average density?   (1) core   (2) crust   (3) mantle   (4) mountains

53. The composition of some meteorites supports the inference that the earth's core is composed of   (1) aluminum and calcium   (2) iron and nickel   (3) silicon and oxygen   (4) magnesium and potassium

Geologic history is recorded in the rocks. Observations of the composition, structure, position, and fossil content of the rock record leads to interpretations of the geologic history of the earth. Carefully study the *Geologic History of New York State at a Glance* and the *Generalized Bedrock Geology of New York State* in the *Earth Science Reference Tables* before reading further in this topic.

RELATIVE DATING OF ROCKS AND EVENTS

Relative dating is the process of obtaining the chronological sequence of rocks or geologic events in an area. The **relative age** of a rock or event is its age as compared to other rocks or events. It should not be confused with the actual age, called the **absolute age,** which refers to the date when the event actually occurred or the rock was formed. Some of the methods of relative dating are described below.

PRINCIPLE OF SUPERPOSITION. In layers of sedimentary rock and some extrusive igneous rocks, the bottom layer is inferred to be the oldest, with each overlying layer being progressively younger, so that the top layer is the youngest. This inference is called the **principle of superposition,** and it may be used as the basis for relative dating. Superposition is due to the original horizontality of deposited sediments. Exceptions to the principle occur in certain types of rock deformation resulting from crustal change. When there are overturns in folds, or when movement along faults has thrust older rock layers over younger layers, the principle of superposition does not hold true (see Figure 13-1 on page 184).

INTRUSIONS AND EXTRUSIONS. When molten rock (magma) invades pre-existing rocks and crystallizes, it forms a body called an **intrusion.** The intrusion is younger than any rock it cuts through (see Figure 13-2 on page 184). When molten rock (lava) flows on the earth's surface and solidifies, it forms a feature called an **extrusion.** The extrusion or lava flow is younger than any rocks beneath it but will be older than any rocks that may later form on top of it.

ROCK STRUCTURAL FEATURES. A rock is older than any fault, *joint,* or fold that appears in the rock. Like a fault, a **joint** is a crack in rocks. However, a joint differs from a fault because there has been no movement of rocks along the crack. If movement (displacement) of the rocks should occur, the joint would become a fault (see Figure 13-3 on page 185).

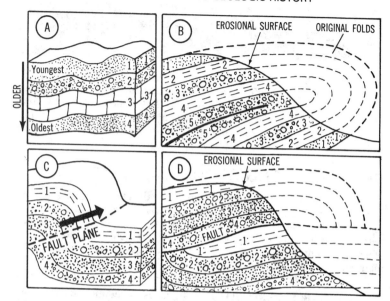

Figure 13-1. The principle of superposition and possible exceptions. (A) Normally, a rock layer is younger than any layers below it. **(B)** Overturned folds can result in an exception to the principle. Layer 1 at lower right is actually a continuation of Layer 1 at upper left. **(C) and (D)** Movement of layers along an overthrust fault can result in an exception to the principle. Layers numbered alike are the same age.

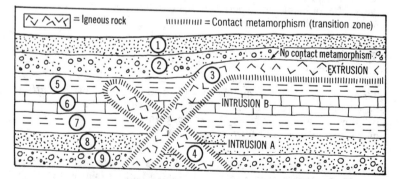

Figure 13-2. Relative ages of intrusions and extrusions. Intrusion A is younger than the layers it intrudes. Intrusion B is younger than A because B cuts A. The layers above the extrusion must be younger than the extrusion because of the absence of contact metamorphism along their common boundary. The rocks are numbered in order of increasing age.

INTERNAL ROCK CHARACTERISTICS. In sedimentary rocks, the sediments are older than the rock itself because they are required to form the rock. In nonsedimentary rocks, individual crystals or fragments are also older than the rock itself.

A **vein** is a mineral deposit, formed from a solution that has filled some crack or permeable zone in rocks. A vein is thus younger than the rock it is in. Mineral cements in sedimentary rocks are also younger than the original sediments in the rock.

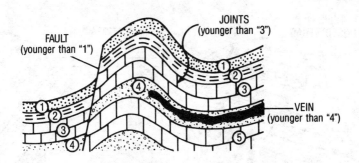

Figure 13-3. Relative ages of structural features of rocks. Folds, faults, joints, and veins are younger than the rocks in which they occur.

CORRELATION

Correlation is the process of showing that rocks or geologic events that occurred in different places are the same age. Correlation is useful in finding mineral deposits, because certain minerals are often found in rocks of a specific age. Correlation is also important in unraveling the sequence of geologic events in an area. Some of the methods of correlation are described below.

CORRELATION BY "WALKING THE OUTCROP." An **outcrop** is local rock, or **bedrock,** that is exposed at the earth's surface. In areas of outcrops, correlation can be accomplished by directly following the continuity of the individual layers or rock formations at the earth's surface; this procedure is called **walking the outcrop** (see Figure 13-4 on page 186). A **formation** is a layer or group of layers of rock and is the basic unit of geologic mapping. The rocks of one formation have similar features, such as rock type, mineral composition, and environment of formation.

CORRELATION BY SIMILARITY IN ROCKS. Where rock formations are separated from one another, they may be tentatively correlated on the basis of similar overall appearance, color, and mineral composition, as indicated in Figure 13-4.

LIMITATIONS OF CORRELATION BY SIMILARITY.   Correlation of rock formations by similarity of the rocks they contain is usually valid over small areas only, and even then may be incorrect. Two rock formations, and even separate parts of the same formation, may be similar and yet be of different ages. For example, Figure 13-5 shows a sedimentary rock formation in which the environment of deposition gradually shifted over a long period of time. Therefore, the rock that formed from the original sediments is much older than the rock that formed in the later stages of deposition, although both parts of the formation are similar in appearance.

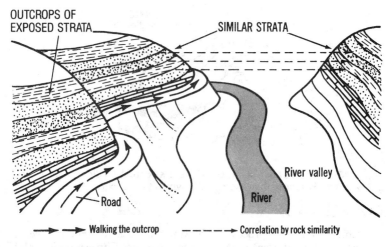

Figure 13-4. Correlation by direct observation. A geologist can follow the strata in an exposed hillside by "walking the outcrop," for example, by walking along the road on the left side of this river valley. The geologist may also be able to correlate strata on opposite sides of the valley by observing similarities in color, texture, and sequence of the rock strata.

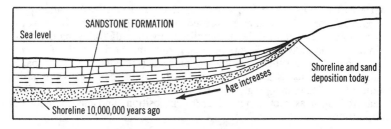

Figure 13-5. Variations in the age of a formation. A sedimentary formation composed of the same type of sediment (for example, sand) may be of different ages in different locations because of movement of the environment of deposition. In this diagram, the sandstone at the left is 10,000,000 years older than the sandstone at the right, even though the formation is continuous and similar in composition.

CORRELATION BY USE OF INDEX FOSSILS. The use of certain *fossils* or groups of fossils is one of the best methods of correlation. A **fossil** is any evidence of former life. With minor exceptions, fossils are found exclusively in sedimentary rocks. Fossils are rarely found in igneous and metamorphic rocks because fossils are usually destroyed by melting and pressure associated with the formation of these nonsedimentary rocks. The fossils used in correlation are called **index fossils.** To be useful as an index fossil, the particular life form must have lived over a large geographic area; that is, it must have a large horizontal distribution throughout rocks formed at the same time. Also, the life form must have lived only a short period of time, thus having a small vertical distribution in the strata in which the fossils occur (see Figure 13-6).

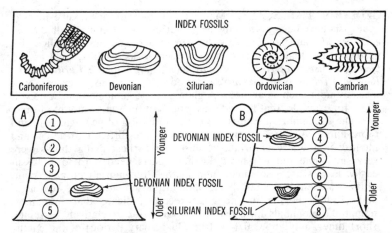

Figure 13-6. Correlation of rocks by means of index fossils. In **A,** a fossil of an organism known to have lived only during the Devonian Period is found in rock layer 4. This layer was therefore deposited during Devonian times. The layers above it are younger and those below it are older. In **B,** a similar fossil is found in one of the layers. This is therefore also a Devonian layer. A second index fossil of the Silurian Period is found in layer 7. This tells us that layers 5 and 6 are not younger than the Devonian and not older than the Silurian. Relative ages of the layers in both formations are indicated by the numbers in order of youngest to oldest.

CORRELATION BY VOLCANIC ASH DEPOSITS. **Volcanic ash** consists of small pieces of extrusive igneous rock that are shot into the air during volcanic eruptions. In large eruptions, volcanic ash is scattered over wide areas of the earth. The ash may then settle among other sediments being deposited in many different regions. If these ash deposits can be detected and identified in rock formations, they are very useful in correlation, because they represent a very small interval of time and because they are widely distributed. They can therefore serve as correlating age markers in rock formations that may be hundreds or thousands of kilometers apart.

GEOLOGIC HISTORY FROM THE ROCK RECORD

FOSSILS AND RELATIVE AGE. Just as the fossils found in rocks can be used to correlate rock strata, they can also be used to place the events in an area in order according to their relative ages. This is possible because the life forms on earth have constantly evolved and some forms existed or were dominant only during specific intervals of geologic time. For example, the rock record shows that the dinosaurs as a group existed in a large interval of geologic time (called the Mesozoic Era), but certain types of dinosaurs existed only during smaller intervals. When these life forms appear as fossils in a given rock formation, they serve to establish the relative age of the formation and to give it a specific place in geologic history.

GEOLOGIC TIME SCALE. Mainly on the basis of such fossil evidence, geologists have been able to divide geologic time into units, called *eons, eras, periods,* and *epochs.* This division of geologic time is called the **geologic time scale** and is often represented as a chart, such as the *Geologic History of New York State at a Glance* in the *Earth Science Reference Tables.* Note that an era, period, or epoch is not an exact unit of measurement such as an hour. A glance at the geologic time scale will show, for example, that no two eras represent the same amount of time.

The Precambrian Eon is the earliest division of geologic time and represents about 88% of the total. Fossils are difficult to detect and identify in the Precambrian, because the earliest living things were very small and did not have the hard parts needed to form fossils easily. Fossils are also uncommon in the Precambrian because many of these ancient rocks have been metamorphosed, melted, or weathered and eroded. Fossils in the rock record indicate that human or humanlike mammals have existed on earth for a relatively extremely short time—only about 0.04% (the Quaternary Period) of the earth's existence.

UNCONFORMITIES. In attempting to read the rock record, geologists often find evidence of buried erosional surfaces called **unconformities** (see Figure 13-7). An unconformity in the rock record of an area indicates that at some time in the geologic history of the area, crustal movement had produced uplift. This uplift had exposed the rocks to weathering and erosion, and part of the rock record was removed. A later subsidence lowered the area, it was covered by water, and new sediments were deposited on the eroded surface, thus producing the unconformity. In most unconformities, there is a lack of parallelism between the older layers below the erosional surface and the younger ones above it as the result of folding or tilting during the uplift or subsidence. However, this is not always the case.

The presence of an unconformity means that some of the layers of the rock record of an area are missing. Although there is a gap in geologic time, an unconformity is useful in relative dating. The rocks above an unconformity are younger than the unconformity, and the rocks below it are considerably older.

Events producing an unconformity with lack of parallelism.

(A) Deposition forms sedimentary rock layers.

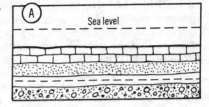

(B) Crustal deformation and uplift occur.

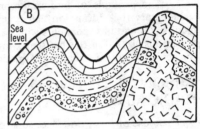

(C) Weathering and erosion remove part of the rock record.

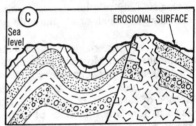

(D) Submergence and new deposition result in an unconformity along the now buried erosional surface.

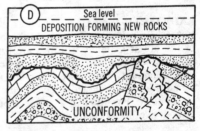

An unconformity in parallel strata.

(E) Uplift, erosion, and submergence have also occurred here, but without deformation. Evidence of a gap in the rock record may be given by widely different ages of the fossils in the layers above and below the unconformity.

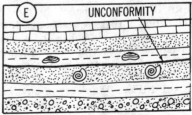

Figure 13-7. Development of an unconformity.

UNIFORMITARIANISM. One of the basic principles geologists use to interpret geologic history is **uniformitarianism,** or uniformity of process, which implies that "the present is the key to the past"; that is, it is assumed that the geologic processes happening today also happened in the past, and that much of the rock record can be interpreted by observing present geologic processes. Uniformitarianism does not mean that different processes could not have happened in the past or that the geologic processes always occurred at the same rates as they do today.

As an example of uniformitarianism, suppose a geologist finds a sedimentary rock formation that is unsorted, has a wide range of sediment sizes, and has sediments with partly rounded shapes with scratches of various sizes. On the basis of knowledge of glacial processes today, the geologist will infer that this formation was the result of past glaciation.

ABSOLUTE DATING OF ROCKS USING RADIOACTIVE DECAY

The use of fossil evidence, superposition, and similar methods gives geologists only relative dates. *Radioactive dating* is their major way of knowing the actual, or absolute, dates for the events of geologic history.

ISOTOPES. An *element* is a substance consisting of atoms that are chemically alike. Most elements exist in several varieties called **isotopes.** The difference between one isotope of an element and another is in the mass of its atoms. For example, the mass of an atom of the most common isotope of carbon is 12 units. This isotope is called carbon-12, to distinguish it from other isotopes of carbon, such as carbon-14, in which the atoms have a mass of 14 units.

RADIOACTIVE DECAY. Almost all the mass of an atom is concentrated in a central region called the *nucleus* (plural, *nuclei*). The nuclei of the atoms of many isotopes are *unstable*. This means that they tend to emit particles and electromagnetic energy, and change to atoms of other elements. This process is called **radioactive decay.** The nucleus that remains after a radioactive decay may also be unstable, and it will decay in its turn. Eventually, a stable isotope (one that does not undergo radioactive decay) is formed.

URANIUM-238. One of the most important radioactive isotopes for the dating of rocks is **uranium-238,** the isotope of uranium whose atoms have a mass of 238 units. The nuclei of its atoms pass through a series of radioactive decays, eventually producing atoms of *lead-206,* a stable isotope of the element lead.

HALF-LIFE. The decay of any individual nucleus is a random event. That is, it may occur at any time. However, among the billions of atoms in any sample of an isotope, a certain definite fraction will decay in a given time. In the next time interval of the same length, the same fraction of the remaining atoms will decay. The time required for half the atoms in a given mass of an isotope to decay is called the **half-life** of the isotope. For any isotope, at the end of one half-life period, half the original atoms will have decayed to other elements, and half will remain unchanged. At the end of the next half-life period, half of these remaining atoms will have decayed, leaving one-fourth of the original atoms unchanged. After a third half-life period, half of these will have decayed, leaving one-eighth of the original atoms unchanged. This halving of the number of unchanged atoms during each successive half-life period continues indefinitely.

The half-life is different for each radioactive isotope, but it is always the same for a given isotope. The half-life of an isotope is not affected by any known environmental factors, such as temperature, pressure, or involvement in chemical reaction. The half-life of an isotope also is not affected by the amount or size of the sample. Each half-life is assumed to have been the same throughout the earth's history. Half-lives vary over a wide range, from fractions of a second to billions of years. (See *Radioactive Decay Data* in the *Earth Science Reference Tables*.)

RADIOACTIVE DATING. The known half-life period of radioactive isotopes can be used to estimate the age of a rock by determining the ratio between the amount of a radioactive isotope and the amount of its decay products in a rock sample. The method is called **radioactive dating.** For example, suppose a rock formed with minerals containing uranium-238 and no lead-206. As time passed, the uranium would slowly change to lead-206 (its stable decay product) at its fixed half-life rate. At the end of one half-life period (4.5 billion years for uranium-238), half the uranium-238 atoms would have changed to atoms of lead-206. In terms of number of atoms, there would be equal amounts of uranium-238 and lead-206, where originally the rock contained no lead-206. If a rock is found today with this 1:1 ratio of uranium-238 atoms to lead-206 atoms, it can be concluded that the rock of formed 4.5 billion years ago. If there is relatively more uranium and less lead, the rock is less than 4.5 billion years old.

The age corresponding to any particular ratio of uranium to lead can be calculated mathematically. The curves in Figure 13-8 on page 192 show how the percentage of uranium-238 decreases and the percentage of lead-206 increases with time in a given rock.

CARBON-14 DATING. Radioactive elements with *long* half-lives (such as uranium-238) are used to date rocks because most rocks have ages measured in hundreds of millions, or even billions, of years. In such long periods, any element with a short half-life would have decayed to such an extent that any remaining amounts would be too small to measure. Some radioactive elements with *short* half-lives, such as *carbon-14* (half-life 5,700 years), are useful for dating remains of *organic materials and rocks of relatively recent origin*. For example, **carbon-14 dating** (also called **radiocarbon dating**) can be used for organic remains up to about 50,000 years in age, which is only the last part of the Pleistocene Epoch ice age.

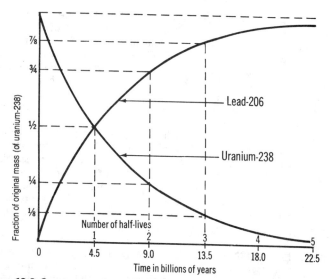

Figure 13-8. Curves showing radioactive decay of uranium-238 and formation of the stable decay product lead-206. After one half-life (4.5 billion years), half the original atoms of uranium have become atoms of lead.

ANCIENT LIFE AND EVOLUTION

VARIATIONS IN FOSSILS AND ENVIRONMENTS.   The fossil record, of mostly sedimentary rocks, shows that a great variety of plants and animals have lived in the past in a great variety of environments. Most of these life forms are now extinct. Comparison of fossils with similar life forms alive today allows geologists to interpret past environments, as shown in Figure 13-9.

The chances of fossilization are very small, so there are probably more types of life forms that have left no evidence of their existence than types of life forms found as fossils.

FOSSILS AND EVOLUTION.   A **species** includes all the life forms that are similar enough to be able to interbreed and produce fertile young. Just by looking around at people, house cats, and dogs, which are three species of life, it can be observed that many *variations* in a species can exist. When graphed, the variations in a species often form bell-shaped curves, as shown in Figure 13-10.

The theory of **organic evolution** assumes that some of the variations within a species give the individuals with those variations a higher chance of surviving and reproducing. If these variations are inheritable, they will be passed on to the offspring, and the favorable variations will be preserved, while unfavorable variations will gradually die out. If this process continues for long periods of time, the accumulated variations may eventually result in a new species—one that can no longer interbreed with the earlier varieties of the species.

The fossil record provides evidence for the theory of evolution. Fossils from adjacent intervals of geologic time show a *gradual transition,* or change, from one species to the next. Recently some scientists have inferred from evidence in the rock record that there are times of rapid organic evolution (punctuated evolution), possibly due to extinctions resulting from collisions between celestial bodies and the earth.

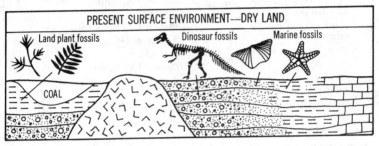

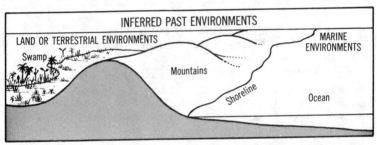

Figure 13-9. Reconstruction of past environments from fossil evidence and rock formations. From the dinosaur and marine fossils and the gradual change in sedimentary rock type as we go further to the right, we infer the former shoreline and ocean. From the types of plant fossils and the occurrence of coal, we infer the former swamp on the left.

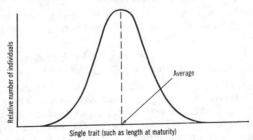

Figure 13-10. Variation in a species trait. In any population of a species, each trait usually has an average value. In most individuals, the trait will be close to the average. In a few, it will be much below average; in a similar number, it will be much above average. If the departure from average in either direction has an advantage for survival and can be inherited, the average will shift in this favorable direction in later generations.

VOCABULARY

relative age	volcanic ash
absolute age	geologic time scale
principle of superposition	unconformity
intrusion	uniformitarianism
extrusion	isotope
joint	radioactive decay
vein	uranium-238
correlation	half-life
outcrop	radioactive dating
bedrock	carbon-14 (radiocarbon)
walking the outcrop	dating
rock formation	species
fossil	organic evolution
index fossil	

QUESTIONS ON TOPIC XIII
INTERPRETING GEOLOGIC HISTORY

Questions in Recent Regents Exams (end of book)

June 1995: 44–50, 91, 92, 94, 95, 97
June 1996: 45–49, 51, 55, 92, 93, 95–97, 100
June 1997: 44–48, 50, 86, 88–90
June 1998: 10, 37, 38, 45–47, 50–52, 91–95, 104, 105

Questions from Earlier Regents Exams

1. According to the *Generalized Geologic Map of New York State,* what is the geologic age of the bedrock found at the surface at 43°30′ N. latitude by 75°00′ longitude?   (1) Devonian   (2) Cambrian   (3) Early Ordovician   (4) Middle Proterozoic

2. Which area in New York State is located on rock formations that contain large amounts of salt deposits?   (1) Syracuse   (2) Long Island   (3) New York City   (4) Old Forge

3. What is the age of the surface bedrock at Niagara Falls, New York? (1) Devonian   (2) Silurian   (3) Ordovician   (4) Cambrian

4. In a certain section of sedimentary rock, fossil dating shows that younger rock layers are on top of older rock layers. This relationship indicates that   (1) the rock layers were formed according to the principle of superposition   (2) the rock layers have been overturned   (3) fossil dating is often inaccurate   (4) the sediments that formed the rock layers were composed of many different minerals

5. The diagram below represents a cross-sectional view of a portion of the earth's crust showing sedimentary rock layers that have not been overturned. The letters identify the specific layers. Which rock layer probably is the oldest?   (1) *A*   (2) *B*   (3) *C*   (4) *D*

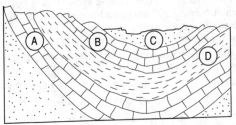

6. The diagram below shows a portion of the earth's crust. What is the relative age of the igneous rock?   (1) It is older than the limestone but younger than the shale.   (2) It is younger than the limestone but older than the shale.   (3) It is older than both the limestone and the shale.   (4) It is younger than both the limestone and the shale.

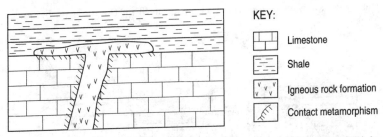

KEY:

▥	Limestone
▤	Shale
ᴠᴠᴠ	Igneous rock formation
◰	Contact metamorphism

7. The diagram below represents an exposed rock outcrop. Which geologic event occurred *last*?   (1) the intrusion of *A*   (2) the fault along line *B*   (3) the fold at *C*   (4) the deposition of gravel at *D*

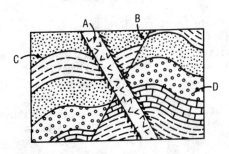

8. The simplest way to correlate exposed rock layers in the same general vicinity when they contain no fossils is by   (1) following vertical veins   (2) radioactive dating   (3) walking the outcrop   (4) tracing a fault

9. The fossil remains of organisms which were once common and widespread but which survived only a short period of geologic time might be especially useful for   (1) correlating sedimentary deposits in places distant from each other   (2) establishing the absolute age of a deposit   (3) tracing the evolution of organisms related to that species   (4) determining some of the factors that led to extinction

10. The diagram below shows a sample of conglomerate rock. The oldest part of this sample is the   (1) conglomerate rock sample   (2) calcite cement   (3) limestone particles   (4) mineral vein

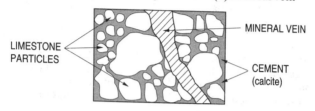

Base your answers to questions 11 through 13 on the diagram below.

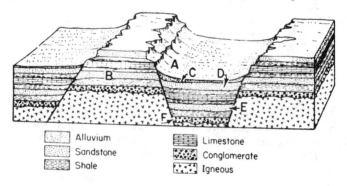

11. The disturbed rock structure shown is probably the result of (1) warping   (2) folding   (3) volcanism   (4) faulting

12. Which is the oldest sedimentary rock in this diagram?   (1) conglomerate   (2) sandstone   (3) shale   (4) limestone

13. Rock layer B is the same age as layer   (1) C   (2) D   (3) E (4) F

14. Below are diagrams of three profile sections showing fossil deposits W, X, Y, and Z, found at widely separated locations. Which would be the best index fossil?   (1) W   (2) X   (3) Y   (4) Z

	LOCALITY A		LOCALITY B	LOCALITY C	
Rock Layer 1	W		W	W	Z
Rock Layer 2	W	Z	Y		Z
Rock Layer 3	W X		X	X	Z

15. The diagram below represents cross sections of three rock outcrops approximately 100 kilometers apart. What would be the best method of correlating the rock layers of each outcrop?   (1) comparing rock types   (2) comparing mineral composition   (3) comparing index fossils   (4) comparing thickness of rock layers

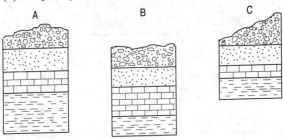

16. Which rock type most likely would contain fossils?   (1) sedimentary rock   (2) metamorphic rock   (3) intrusive igneous rock   (4) extrusive igneous rock

17. Why can layers of volcanic ash found between other rock layers often serve as good geologic time markers?   (1) Volcanic ash usually occurs in narrow bands around volcanoes.   (2) Volcanic ash usually contains index fossils.   (3) Volcanic ash usually contains the radioactive isotope carbon-14.   (4) Volcanic ash usually is rapidly deposited over a large area.

18. According to the *Earth Science Reference Tables*, which rock is most likely the oldest?   (1) conglomerate containing the tusk of a mastodon   (2) shale containing trilobite fossils   (3) sandstone containing fossils of flowering plants   (4) siltstone containing dinosaur footprints

19. The best basis for concluding that a certain layer of shale rock in New York State was deposited at the same time as one in California is that both   (1) are the same distance below the surface   (2) contain similar fossil remains   (3) are sedimentary rocks   (4) have the same chemical composition

20. Which rock layer is *not* found in the rock record of New York State? (1) Devonian   (2) Silurian   (3) Ordovician   (4) Permian

21. During which time was the majority of the exposed bedrock in New York State deposited?   (1) Precambrian   (2) Mesozoic   (3) Cenozoic   (4) Paleozoic

22. Rocks containing fossils of the earliest land plants could most likely be found in New York State bedrock near   (1) Syracuse   (2) Oswego   (3) Ithaca   (4) Old Forge

23. Approximately how long ago were the Taconic Mountains uplifted?   (1) 540 million years ago   (2) 440 million years ago   (3) 310 million years ago   (4) 120 million years ago

24. According to the Geologic Time Scale in the *Earth Science Reference Tables*, what is the estimated age of the earth as a planet in millions of years?   (1) 540   (2) 4000   (3) 4600   (4) 5000

25. The Geologic Time Scale has been subdivided into a number of time units called periods on the basis of   (1) fossil evidence   (2) rock thicknesses   (3) rock types   (4) radioactive dating

26. Which line is the best representation of the relative duration of each of the geologic time intervals?

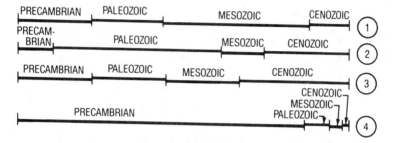

Base your answers to questions 27 through 31 on the diagram below.

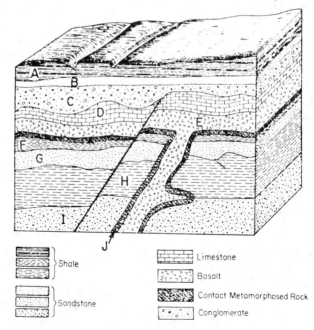

27. Which is the most recently formed rock?   (1) A   (2) B   (3) C   (4) D

28. An unconformity is located between   (1) A and B   (2) B and C   (3) C and D   (4) F and G

29. The conclusion that the limestone layer (D) is younger than the basalt layer (E) is supported by evidence of   (1) faulting   (2) contact metamorphism   (3) igneous intrusion   (4) fossils

30. The last event before faulting occurred was formation of   (1) C   (2) D   (3) E   (4) J

31. Which must have preceded the formation of layer D?   (1) faulting   (2) submergence   (3) intrusion   (4) uplift

32. An unconformity between two sedimentary layers is most likely produced by   (1) the deposition of gravel followed by the deposition of sand and silt   (2) continuous sedimentation in a deep basin over a long period   (3) uplift followed by extensive erosion, submergence, and deposition   (4) a period of extrusive vulcanism followed by another period of extrusive vulcanism

33. The diagram below shows a cross-sectional view of part of the earth's crust. What does the unconformity (buried erosional surface) at line *XY* represent?   (1) an area of contact metamorphism   (2) a time gap in the rock record of the area   (3) proof that no deposition occurred between the Cambrian and Carboniferous periods   (4) overturning of the Cambrian and Carboniferous rock layers

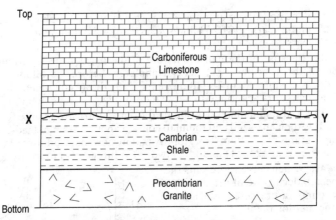

34. The physical and chemical conditions which long ago produced changes on the earth's surface are still producing changes. This statement is one way of stating the principle of   (1) catastrophism   (2) diastrophism   (3) uniformitarianism   (4) isostasy

35. The age of the earth is most accurately estimated from   (1) the salinity of the oceans   (2) the thickness of sedimentary rock   (3) studies of fossils   (4) radioactive dating of rock masses

36. "The present is the key to the past." This statement is best interpreted as meaning that   (1) rocks now exposed were formed in the past   (2) the processes by which rocks are formed today have been going on throughout the earth's history   (3) activity occurring now will cease in the near future   (4) animals living now are unlike any that lived in the geologic past

Base your answers to questions 37 through 41 on the graph below, which shows the radioactive decay curve for uranium to lead.

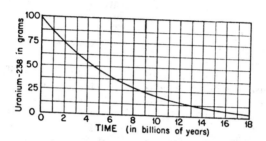

37. The half-life of uranium-238 as shown on this graph is closest to (1) 2.5 billion years   (2) 3.5 billion years   (3) 4.5 billion years   (4) 5.5 billion years

38. Because of the half-life nature of the decay process, uranium-238 will theoretically   (1) decay completely before 4.5 billion years   (2) decay completely in 18 billion years   (3) decay completely in 200 billion years   (4) continue to decay indefinitely

39. A uranium mineral is obtained from an intrusive granite formation. It is then analyzed and found to contain about 1 gram of lead-206 to every 3 grams of uranium-238. Approximately how many billions of years old is the granite?   (1) one   (2) two   (3) three   (4) four

40. Which curve represents the expected decay rate of uranium-238 if its temperature were raised almost to the melting point?

41. If the decay curve for the element carbon-14 were plotted, the general shape of the curve would be

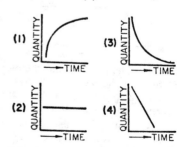

42. The half-life of a particular radioactive substance   (1) decreases as pressure on it increases   (2) decreases as its mass decreases   (3) increases as the temperature increases   (4) is independent of mass, temperature, and pressure

43. If a specimen of a given radioactive substance is reduced in size, its half-life   (1) decreases   (2) increases   (3) remains the same

44. Which radioactive substance shown on the graph below has the longest half-life?   (1) A   (2) B   (3) C   (4) D

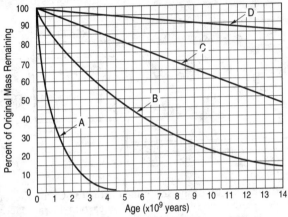

45. Why is carbon-14 *not* usually used to accurately date objects more than 50,000 years old?   (1) Carbon-14 has a relatively short half-life and too little carbon-14 is left after 50,000 years.   (2) Carbon-14 has a relatively long half-life and not enough carbon-14 has decayed after 50,000 years.   (3) Carbon-14 has been introduced as an impurity in most materials older than 50,000 years.   (4) Carbon-14 has only existed on earth during the last 50,000 years.

46. The element K^{40} radioactively decays to Ar^{40}. The diagram at the right shows a model of the relative amounts of K^{40} and Ar^{40} remaining after one half-life. Which diagram below best illustrates the relative amounts of K^{40} and Ar^{40} remaining after two half-lives?

47. Uranium-238 is used to date the age of the earth rather than carbon-14 because uranium-238   (1) was more abundant when the earth formed   (2) has a longer half-life   (3) decays at a constant rate   (4) is easier to collect and test

48. The diagram below represents a section of the earth's crust. The symbols in the diagram indicate the location on a horizontal surface of certain fossils that formed during the Carboniferous Period. For what purpose would the fossil information on the map be most useful?

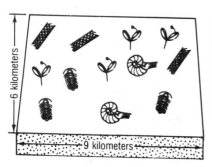

KEY TO FOSSIL SYMBOLS	
Land Fossils	Marine Fossils
scale tree	trilobite
seed fern	mollusk

(1) to find the location of the shoreline during the Carboniferous Period
(2) to measure the age of the bedrock by carbon-14 radioactive dating
(3) to provide evidence of the evolution of humans   (4) to indicate the extent of folding that occurred during the Devonian Period

49. Shark and coral fossils are found in the rock record of certain land areas. What does the presence of these fossils indicate about those areas?   (1) They have undergone glacial deposition.   (2) They were once covered by thick vegetation.   (3) They have undergone intense metamorphism.   (4) They were once covered by shallow seas.

50. Earth scientists studied fossils of a certain type of plant. They noted slight differences in the plant throughout geologic time. What inference is best made from this evidence?   (1) When the environment changed, this type of plant also changed, allowing it to survive. (2) When uplifting occurred, the fossils of this type of plant were deformed.   (3) The processes which form fossils today differ from those of the past.   (4) The fossils have changed as a result of weathering and erosion.

51. The changes observed in the fossil record from the Precambrian era to the Cenozoic Era best provide evidence of   (1) sublimation (2) radioactive decay   (3) evolution   (4) planetary motion

52. Which statement is false?   (1) Two members of one species must be exactly alike.   (2) The members of a species can vary in physical features.   (3) The members of a species can interbreed and produce fertile young.   (4) A species may exist on earth for millions of years.

TOPIC XIV — Landscape Development and Environmental Change

Landscapes (topography) are the features of the earth's surface at the interface between the atmosphere, hydrosphere, and lithosphere (crust). Some of the characteristics of landscapes are the slope of the land, shape of the surface features, stream drainage patterns, stream slope, and soil characteristics. A **stream drainage pattern** is the shape of the aerial view of the stream courses in an area (see Figure 14-6 on pages 210–211).

MEASURING LANDSCAPE CHARACTERISTICS

The shape and slope of the land, stream drainage patterns, and some soil features can be measured using actual observations or models such as contour maps, aerial photographs, satellite images, and the types of maps and diagrams used in this topic. The use of these tools has shown that landscape features such as hills, slopes, and stream drainage patterns have distinctive shapes by which they can be identified.

GRADIENT AND PROFILES. *Slope,* or *gradient,* a measurable characteristic of the land and streams, is described in Topic III. *Profiles,* also described in Topic III, are useful to show the shape and slope of the earth's surface and streams, and the thickness and development of soil horizons.

MOUNTAINS, PLATEAUS, AND PLAINS. On the basis of gradient, elevation, and rock structure, landscapes are divided into three major types—*mountains, plateaus,* and *plains* (see Figure 14-1).

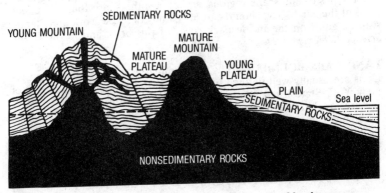

Figure 14-1. Profiles and structures of the major types of landscapes.

A **mountain** is an area of high elevation compared to the surrounding area or sea level, which usually has many changes in slope as well as regions of steep gradient. Internally, mountains are characterized by distorted rock structures such as faults, folds, and tilted rocks, and are often composed of much nonsedimentary rock.

A **plateau** is also an area of high elevation, but it has undistorted horizontal rock structure and often a more level slope or gradient than mountains. A plateau may have steep slopes where streams have cut valleys, such as at the Grand Canyon in the Colorado Plateau.

A **plain** has a generally level surface with little change in slope and has a low elevation. Plains usually have horizontal rock structure unless they are remnants of old mountain areas that have been leveled by erosion and weathering.

SOIL ASSOCIATIONS. Soils differ in composition, particle size, structure, permeability, porosity, fertility, and degree of horizon development. Soils of similar characteristics are grouped together as a **soil association,** which is similar to a formation in rocks. The boundaries of different associations in an area are often indicated on maps.

LANDSCAPE REGIONS

Landscape characteristics (amount of slope, elevation, bedrock structure, stream drainage patterns, and soil characteristics) appear to occur in combinations that form identifiable areas called **landscape regions** or physiographic provinces. For example, the combination of high elevation, steep slopes, thin soils, and trellis, annular, and radial stream patterns is common in mountain landscape regions (see Figure 14-6).

CONTINENTAL LANDSCAPE REGIONS. Any continental land mass has several landscape regions that can be identified. The general landscape regions of the continental United States are shown in Figure 14-2. These types of landscape regions, as well as others, are found in many parts of the earth. You should carefully study the New York State landscape regions on the inside back cover of this book and in the *Earth Science Reference Tables.*

LANDSCAPE BOUNDARIES. The boundaries between landscape regions are usually well defined, or distinct, as illustrated on the maps of New York State listed above. Landscape boundaries usually consist of features or characteristics of the landscape that have been brought about by changes in the structure of rocks. Such features and characteristics include the edges of mountains, cliffs, changes in type or amount of slope, or river courses that follow the direction of structural changes.

FACTORS OF LANDSCAPE DEVELOPMENT

The process of landscape development is very complex and involves many factors that may or may not be operating at a particular time and

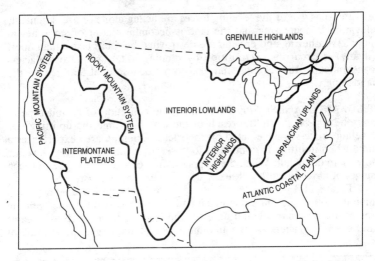

Figure 14-2. Major landscape regions of the contiguous United States.

place. Therefore, generalizations about landscape development may not apply in any one specific area.

UPLIFTING AND LEVELING FORCES. There are two groups of forces that operate to form and change landscapes—the *uplifting* (or constructional) forces and the *leveling* (or destructional) forces.

Uplifting forces originate beneath or within the earth's crust and displace rock material to raise the land, build mountains, and cause continental growth. The uplifting forces include volcanic action, isostasy, earthquakes, and many plate tectonic events.

Most **leveling forces** operate on the earth's surface. They break down the rocks of land masses and transport material along the earth's surface from higher to lower elevations under the force of gravity, thus tending to level out the land. The leveling forces include weathering, erosion, deposition, and subsidence.

Leveling forces (such as erosion or weathering) are always at work in all areas of the earth's solid surface, but uplifting may or may not be present at the same time. When both groups or forces are present, the landscape will be uplifted or leveled, depending on which forces are dominant (operating at a faster rate).

If uplifting is dominant, level land of low elevation may be changed into mountains with steep hillslopes and high elevations, stream drainage patterns may be altered as the streams begin to erode new landscapes, and wind patterns may change as the mountains act as barriers to the wind. Uplifting forces are usually dominant in plate boundary regions of earthquakes and volcanic action, where mountain building is going on. The landscape reflects the effects of those forces.

In areas where the leveling forces are acting alone or are dominant, elevation is decreasing and the land is becoming flatter or smoother in slope. Weathering and erosion are destroying the rocks faster than uplift (if any) is occurring. Deposition by running water, wind, and glaciers contributes to the leveling and also forms depositional features on the landscape.

TIME AND LANDSCAPE STAGES. The condition of a landscape at any one time is partly the result of the length of time the uplifting and leveling processes have acted on the rocks. Often an area experiences a time when uplifting is dominant, followed by a long time when leveling forces are dominant. When this happens, an area's landscape goes through **stages of development,** with each stage having characteristic conditions.

Figure 14-3 is a model of the stages of landscape development in a mountainous region in a humid climate. In stage A (youth), the uplifting forces are dominant. Uplifting gives the leveling forces such as streams and possibly glaciers a large amount of potential energy, and the rate of

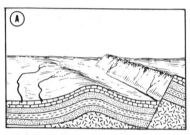 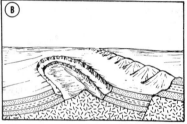

Youth. Uplifting forces are dominant, causing folding and faulting, and forming mountains with high elevations and steep slopes.

Maturity. Leveling forces are dominant, creating a rugged landscape with lower elevations.

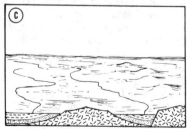 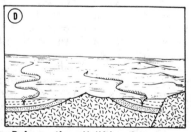

Old age. Leveling forces are still dominant, but less effective because low elevations and gentle slopes provide little potential energy.

Rejuvenation. Uplifting forces are again dominant. As elevations increase, streams acquire more potential energy and form new valleys and steeper slopes.

Figure 14-3. Stages of landscape development.

leveling is great. As the uplifting forces decrease in effect, the destructional forces carve away the land, creating a rugged landscape of lower elevation characteristic of stage B (maturity). If the leveling forces continue to be dominant for millions of years, the landscape will become progressively smoother in slope and lower in elevation. When the former mountain region is smooth enough to resemble a plain, it is considered to be in stage C (old age). In stage C the leveling forces are usually less active than in earlier stages because of flat slopes and low elevation, so that streams have small potential energy and small gradient.

DYNAMIC EQUILIBRIUM IN LANDSCAPES. Although landscape development is a continuous process leading to gradual change in the landscape, at any one time the landscape condition reflects a state of balance among many environmental factors. This is a condition of **dynamic equilibrium.** If a change occurs in any of the factors, a change in the landscape features will occur until a new equilibrium is established (Stage D in Figure 14-3).

FACTORS OF
LANDSCAPE DEVELOPMENT—CLIMATE

The rate of development and the characteristics of the landscape of an area are greatly influenced by the temperature and moisture conditions of the area. Any change in an area's climate will alter the rate of development and characteristics of the landscape. On the other hand, changes in landscape sometimes cause changes in climate. For example, the growth of a new mountain range will block air mass movement and change moisture and temperature patterns, thereby changing the climate of affected areas. In turn, more landscape changes will occur in response to the climatic change. The effect of mountain ranges on climate, called the *orographic effect,* is described in Topic VIII, pages 117-118.

HILLSLOPES IN ARID AND HUMID CLIMATES. The steepness of hillslopes is partly affected by the balance between production of sediments by weathering and the removal of these sediments by erosion. In

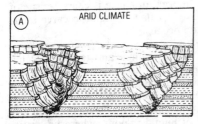
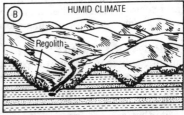

Figure 14-4. Landscape development of a plateau as affected by climate. The rock formations are assumed to be similar in both diagrams. **A** illustrates the steep slopes characteristic of an arid climate. **B** illustrates the smoother, more rounded landscape features of a humid climate, such as the Catskills of New York State.

arid climates there is little vegetation to hold sediments in place on slopes, or as deposits at the base of slopes, so wind and running water can rapidly carry the sediments away. This rapid removal of sediments causes many arid regions to be characterized by the steep slopes and sharp landscape features common in the southwestern United States. Arid regions are also characterized by sand dunes and by bedrock sculptured by wind-blown sand.

In humid climates, sediments on and at the bottom of hillslopes are better held in place by vegetation; therefore, areas with humid climates are characterized by the smoother and more rounded landscape features common in the eastern United States. The greater rate of chemical weathering of bedrock in humid climates contributes to their more rounded landscapes.

GLACIATION AND LANDSCAPES. In climates where glaciers exist or existed in recent times (such as during the Pleistocene—see the *Geologic History of New York State at a Glance* in the *Earth Science Reference Tables*), the landscape will show much evidence of glacial erosion and deposition. Some of the landscape features of glaciation are (1) mountaintops and steep slopes without much soil, (2) transported soil covering large areas, (3) soil with a wide range of particle sizes even at the surface, (4) wide valleys with U-shaped profiles (as compared with a stream's V-shaped profile), (5) many lakes, (6) disrupted stream drainage patterns, (7) many small hills composed of sediment, (8) polished and scratched bedrock, and (9) the features of the glaciers themselves where the glaciers still exist.

STREAMS AND CLIMATE. Some of the characteristics of streams are controlled by climate. In arid regions, most streams, being temporary, are without water for much of the time; streams in humid regions are permanent, having discharge most or all of the time.

Internal drainage is common in arid regions. Internal drainage occurs when water is channeled into basins that are not connected by streams to the oceans. Examples of such basins include the Great Salt Lake and others in Death Valley, California. Internal drainage is characteristic of arid regions because the streams have not had the time to carve interconnecting valleys for drainage to the oceans.

SOILS AND CLIMATE. One of the most important factors in determining soil characteristics is climate. Soils in arid regions are often thin or nonexistent, because there is little or no vegetation to make and hold the soil. The soil that does exist is often very sandy, because smaller sediments have been blown away. Arid soils often contain many mineral salts that are not found in the soil of humid regions, where infiltration and runoff dissolve and carry the salts away. In humid regions, the soils tend to be thicker and more acidic; they are also higher in organic content and thus darker in color. If the climate is very hot and humid, the soil will be infertile because chemical weathering and infiltration rapidly remove organic and mineral nutrients, leaving the soil deficient in these components.

FACTORS OF
LANDSCAPE DEVELOPMENT—BEDROCK

The composition and structural features of the bedrock are major factors in the rate of development and the characteristics of landscapes.

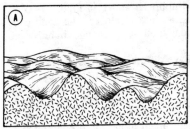

Homogeneous nonsedimentary bedrock. The landscape consists of a random distribution of rounded hills.

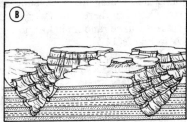

Horizontal sedimentary bedrock (plateau). The landscape has generally uniform elevation, with steep-sided valleys cut by streams.

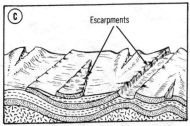

Folded strata of varying resistance. The landscape consists of roughly parallel ridges, with steep-sided escarpments of resistant rock.

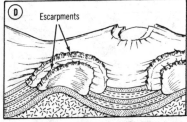

Domed structure. Landscape features resemble **C**, but ridges and escarpments have a generally circular arrangement.

Fault block mountains. The landscape features consist of ridges of varying elevation, with steep slopes along the fault plantes.

Complex bedrock structure. The landscape consists of a great variety of elevations and slopes.

Figure 14-5. Effect of bedrock structure on landscape features.

ROCK RESISTANCE AND HILLSLOPES. In any one climate, different rock types and formations have varying degrees of **rock resistance,** or resistance to weathering and erosion. If all the rocks in an area exposed to the surface have about the same resistance, the landscape features will be controlled by structural features, such as faults and joints, and the agents of leveling and uplift (E and F in Figure 14-5 on page 209). If there are no special structural features, the landscape features will be random in location and without rapid changes in hillslopes (A in Figure 14-5).

If rocks exposed at the surface have different degrees of resistance, the different rocks will weather and erode at different rates. For example, in sedimentary strata, sandstone layers will generally be more resistant than shale and will weather and erode more slowly. The result will be marked differences in slope between the layers of different resistance, often with a steep slope called an **escarpment** along the eroded edge of the more resistant layer (B, C, and D in Figure 14-5).

Figure 14-6. Effect of bedrock structure on stream drainage patterns. Under each sketch of the landscape there is a corresponding contour map showing the stream drainage.

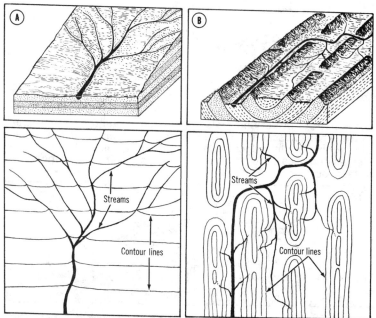

Random or dendritic drainage. This pattern is characteristic of horizontal sedimentary rocks with little difference in rock resistance.

Trellis or block drainage. This is observed in folded rocks with much difference in resistance, and also in faulted or jointed rock.

STRUCTURAL FEATURES AND HILLSLOPES. Often the types of **rock structure,** such as horizontality, folds, faults, and joints, have a major effect on hillslopes. A study of the diagrams in Figure 14-5 will show the influence of horizontal rocks, folded rocks, and faulted rocks. Generally, the greater the variety of structural features found in an area, the more varied the changes in hillslope and types of landscape features (F in Figure 14-5).

STREAMS AND BEDROCK CHARACTERISTICS. The direction, pattern, and gradient of streams are often directly related to the resistance and structure of underlying bedrock. In horizontal rocks with little difference in resistance, the streams will develop a random pattern and have few rapid changes in gradient, as shown in A of Figure 14-6. When the rocks in an area have varying degrees of resistance, or there are tilted, folded, faulted, or jointed rocks, the patterns and directions of the streams will be at least partly if not totally controlled by these rock features, as shown in B, C, and D. In such cases there is often sharp change in gradient, as indicated by falls and rapids.

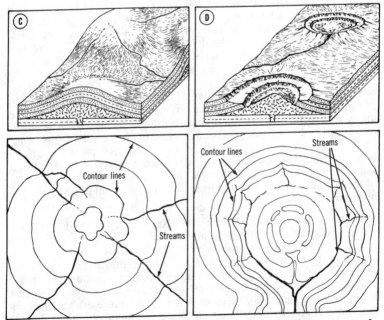

Radial drainage. This pattern occurs in an area of domed structure with little difference in rock resistance.

Annular drainage. This pattern of concentric circles is found in areas of domed structure with much difference in rock resistance.

SOILS AND ROCK COMPOSITION. The different compositions of the soil associations are due in part to differences in the underlying bedrock or in transported material from which the soil formed. A soil formed from a sedimentary rock such as sandstone will be much different in composition from a soil formed from an igneous rock such as basalt, because of major differences in the mineral and chemical composition of the rocks. A soil association map and a geologic map (see *Generalized Bedrock Geology of New York State* in the *Earth Science Reference Tables*) will often show a close correlation between soil and rock type.

PEOPLE AND ENVIRONMENTAL CHANGE

People have greatly affected their environment, including the landscape. They have cut down forests and plowed up the land, allowing soil to be carried away or *denuded;* they have carved up or smoothed out the land for mining and construction of roads, buildings, and airports; they have polluted the land with their discards; they have added chemicals to the land, air, and water which aid weathering, and they have even made new land by filling in lakes and parts of the ocean with their garbage. Thus people are a major factor in landscape development.

LANDSCAPE POLLUTION AND POPULATION. Figure 14-7 is a graph showing that the human population has been increasing at a rapid or exponential rate in recent times. Landscape pollution is generally greatest in areas of high population density. This is so because it is there that people erode and deposit material in concentrations large enough to adversely affect their lives and the life forms and landscape.

Modern people can cause rapid changes in their environment because of **technology**—the application of scientific discoveries to the methods of producing goods and services. Catastrophic events may occur as the environment reacts to the changes caused by technology. The effects of atomic bombs, deforestation of thousands of acres a year, the plowing-up of thousands of acres of grassland a year, the use of millions of tons of pesticides a year, and similar changes cause rapid responses in wildlife, soil and stream characteristics, and rates of landscape evolution. Many of these changes have resulted in the extinction of wildlife and the loss of useful landscapes.

ATMOSPHERIC POLLUTION AND LANDSCAPE CHANGE. The addition of liquid and solid aerosols to the atmosphere by people can increase reflection and scattering of insolation. This may result in less energy at the surface of the earth for natural landscape-producing processes such as weathering. On the other hand, the addition of carbon dioxide and water vapor to the atmosphere can increase the greenhouse effect and thus cause an increase of energy at the earth's surface for landscape change. Many industrial and community activities add substances to the atmosphere that cause areas to have greater amounts of cloud coverage than would naturally occur. Increased cloud coverage affects precipitation and temperatures and thereby can alter the type and rate of landscape development. The addition of sulfur compounds to the atmosphere, largely by burning fossil fuels and mineral refining,

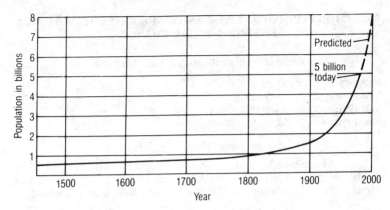

Figure 14-7. Estimated human population of the world in recent centuries (projected to the year 2000).

has significantly increased the acid content of precipitation. This acid precipitation (acid rain) has increased the chemical weathering of rocks, regolith, and vegetation, resulting in changed landscapes.

ENVIRONMENTAL CONSERVATION. Resources, such as soil for agriculture, land for homesites, pure water for consumption and recreation, and clean air for biologic activity, can be conserved by careful planning and by the control of environmental pollutants.

Some of the goals of environmental planning are elimination of landscape pollution and a net elimination of *denudation*. **Denudation** is the loss or removal of regolith (soil and sediments) and life forms, such as trees, due to leveling forces. Often denudation is due to or accelerated by human activities. Another goal is the reclaiming of landscapes that have been made unusable by misuse.

For successful programs in conservation of the environment, there must be education to make people aware of the problems and to change attitudes that lead to misuse of the environment. Finally, people must *act* to stop further misuse and to repair former destruction.

VOCABULARY

landscape	leveling forces
stream drainage pattern	stages of landscape development
mountain	dynamic equilibrium in landscape
plateau	rock resistance
plain	escarpment
soil association	rock structure
landscape region	technology
uplifting forces	denudation

QUESTIONS ON TOPIC XIV—LANDSCAPE DEVELOPMENT AND ENVIRONMENTAL CHANGE

Questions in Recent Regents Exams (end of book)

June 1995: 51–55, 93, 98–100, 105
June 1996: 33, 50, 52–54, 99
June 1997: 5, 26, 39, 49, 51–55, 96, 98
June 1998: 31, 36, 48, 53–55, 96–100

Questions from Earlier Regents Exams

1. Which New York State landscape region is best represented by the block diagram below?   (1) Allegheny Plateau   (2) Adirondack Mountains   (3) Atlantic Coastal Plain   (4) Erie-Ontario Lowlands

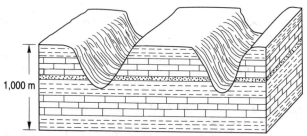

2. Although the Adirondacks are classified as a mountain landscape, the Catskills are classified as a plateau landscape because of a major difference in their   (1) amounts of rainfall   (2) bedrock structure   (3) index fossils   (4) glacial deposits

3. Characteristics such as composition, porosity, permeability, and particle size are used to describe different types of   (1) hillslopes   (2) stream drainage patterns   (3) soils   (4) landscapes

4. The major landscape regions of the United States are identified chiefly on the basis of   (1) similar surface characteristics   (2) similar climatic conditions   (3) nearness to major mountain regions   (4) nearness to continental boundaries

5. According to the *Earth Science Reference Tables*, the location 44° N latitude and 74°30′ W longitude is found in which New York State landscape region?   (1) Hudson-Mohawk Lowlands   (2) St. Lawrence Lowlands   (3) Adirondack Mountains   (4) the Catskills

6. The primary reason that several landscape regions have formed in New York State is that the various regions of the State have different (1) climates   (2) latitudes   (3) soil characteristics   (4) bedrock characteristics

7. According to the *Earth Science Reference Tables*, which New York State landscape region has the lowest elevation, the most nearly level land surface, and is composed primarily of Cretaceous through Pleistocene unconsolidated sediments?   (1) the Hudson-Mohawk Lowlands

(2) the Atlantic Coastal Plain   (3) the Champlain Lowlands   (4) the Erie-Ontario Lowlands

8. Boundaries between landscape regions in New York State are best described as being   (1) straight lines running north and south   (2) invisible on the surface   (3) usually well defined   (4) generally unchanging

9. The boundaries between landscape regions are usually indicated by sharp changes in   (1) bedrock structure and elevation   (2) weathering rate and method of deposition   (3) soil associations and geologic age   (4) stream discharge rate and direction of flow

10. Landscape regions in which leveling forces are dominant over uplifting forces are often characterized by   (1) volcanoes   (2) mountain building   (3) low elevations and gentle slopes   (4) high elevations and steep slopes

11. When most landscape regions are uplifted, the amount of weathering and erosion that occurs will generally   (1) decrease   (2) increase   (3) remain the same

12. Which change would be occurring in a landscape region where uplifting forces are dominant over leveling forces?   (1) topographic features that are becoming smoother with time   (2) a state of dynamic equilibrium existing with time   (3) streams that are decreasing in velocity with time   (4) hillslopes that are increasing in steepness with time

13. Which cross-sectional diagram below best represents a landscape region that resulted from faulting?

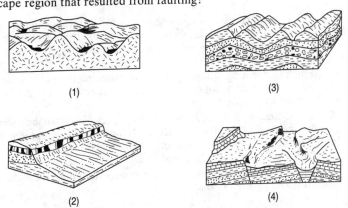

(1)

(3)

(2)

(4)

14. Which characteristics of a landscape region would provide the best information about the stage of development of the landscape?   (1) the age and fossil content of the bedrock   (2) the type of hillslopes and the stream patterns   (3) the amount of precipitation and the potential evapotranspiration   (4) the type of vegetation and the vegetation's growth rate

15. The three diagrams below represent the same location at different times. What is the best explanation for the differences in appearance of this location?

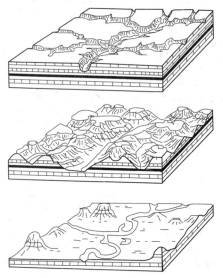

(1) This region was exposed to leveling forces for different lengths of time.   (2) This region has different underlying bedrock structures. (3) This region developed in different climate regions.   (4) This region has the same type of underlying rock layers.

16. Which factor is most important in determining the evolution of a landscape?   (1) surface topography   (2) plant cover   (3) climate (4) development of drainage

17. A landscape is characterized by much transported soil, scratched rock surfaces, and wide U-shaped valleys. The development of this landscape is probably the result of   (1) stream action   (2) wind erosion   (3) uplifting   (4) glaciation

18. Which diagram best represents a cross section of a valley which was glaciated and then eroded by a stream?

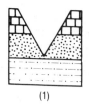

(1)

(2)

(3)

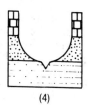

(4)

19. In the cross section of the hill shown below, which rock units are probably most resistant to weathering?

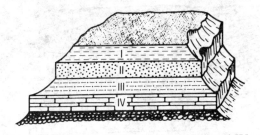

(1) I and II   (2) II and III   (3) I and III   (4) II and IV

20. How was the valley shown in the diagram below most likely formed?

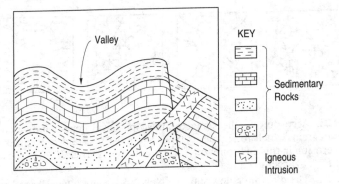

(1) by the deposition of sediments   (2) by the extrusion of igneous material   (3) by the faulting of rock layers   (4) by the folding of rock layers

21. Which of the following stream patterns is most characteristic of horizontal rock structure?

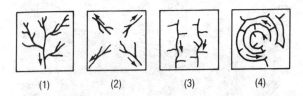

Base your answers to questions 22 and 23 on the diagram below of a section of the earth's crust.

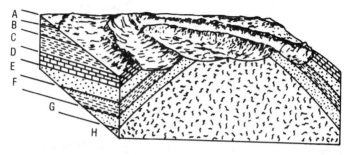

22. Which kind of stream pattern would most likely be found on the type of landscape shown in the diagram?

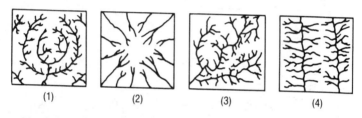

(1)   (2)   (3)   (4)

23. According to the surface landscape development indicated in the diagram, which rock type or types are most resistant to weathering and erosion in this environment?   (1) rock E (2) rock H (3) rocks C and F (4) rocks D and G

24. The diagram below shows a New York State highway road cut.

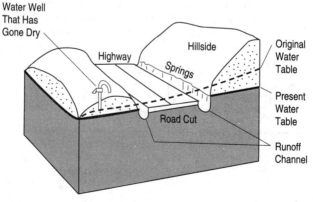

The water well has probably gone dry because   (1) crustal uplift has altered the landscape   (2) human activities have altered the landscape   (3) the climate has become more humid   (4) the bedrock porosity has changed

25. Humans can cause rapid changes in the environment, which may produce catastrophic events. Which statement below is the best example of this concept?   (1) Mountainside highway construction causes a landslide.   (2) Lightning causes a forest fire.   (3) Shifting crustal plates cause an earthquake.   (4) Changing seasonal winds cause flooding in an area.

26. Which graph best represents human population growth?

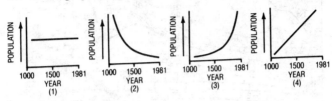

27. As population increases, the amount of alteration of the landscape by human activity will most likely   (1) decrease   (2) increase   (3) remain the same

28. The addition of pollutants to the atmosphere may change the rate of energy absorption and radiation. This could result in   (1) a renewal of volcanic activity   (2) the formation of a new fault zone   (3) a landscape-modifying climate change   (4) a change in the times of high and low tides

Additional Questions

1. Models of hillslopes and landform shapes can be made with the use of   (1) isobars   (2) contour lines   (3) isotherms   (4) parallels

2. The different types of soil found at the earth's surface are classified as   (1) soil profiles   (2) soil horizons   (3) soil associations   (4) soil formations

3. If a landscape is in dynamic equilibrium and uplift occurs, (1) there can never again be dynamic equilibrium in the area   (2) the dynamic equilibrium will remain the same   (3) a new dynamic equilibrium will be established   (4) a new landscape will form without dynamic equilibrium

4. If a new mountain range were to form where the Mississippi River is today, the landscape east of this new mountain range would become more angular and less rounded because the climate would become   (1) drier and warmer   (2) drier and cooler   (3) moister and warmer   (4) moister and cooler

5. Which of the following would *not* explain why soils are different in separate locations?   (1) different climates   (2) different compositions of the rocks the soils formed from   (3) different amounts of time for soil formation   (4) different ages of the rocks the soil formed from

Refer to the diagram below to answer questions 6 through 11.

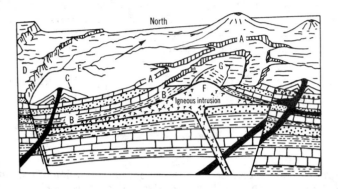

6. The climate of this area is most likely   (1) arid   (2) humid
(3) hot   (4) cold

7. The cliffs around location *A* are most directly the result of
(1) igneous intrusion   (2) resistant rock layers   (3) the different ages
of the rock   (4) movement along faults

8. What can be said about rock layer *B* compared to rock layer *F*?
(1) *B* is younger   (2) *B* is more resistant   (3) *B* is less resistant   (4) *B*
is lower because of its environment of deposition

9. How would you classify the geologic forces producing the landscapes at locations *C* and *G*? (1) At both places uplifting forces are
dominant.   (2) At both places leveling forces are dominant.   (3) At *C*
uplifting is dominant, and at *G* leveling is dominant.   (4) At *C* leveling
is dominant, and at *G* uplifting is dominant.

10. What immediate effects would increased rainfall have on
the rate of leveling of this landscape?   (1) increase   (2) decrease
(3) remain the same

11. The primary reason for the cliff at *D* is   (1) volcanic action
(2) resistant rock layers   (3) movement along a fault   (4) joints in the
rock layers

12. What is the most practical way to control pollution?   (1) slow-
ing down the rate of technological growth   (2) decreasing the population   (3) careful planning of environmental usage   (4) more govern-
ment control of industry

13. The most practical way individuals can help fight pollution is by
(1) becoming aware of the problems and thus changing their attitudes
and actions   (2) recycling paper   (3) using fewer products produced by
technology   (4) changing their life-styles to conform to that of
nineteenth-century rural America

Glossary

absolute age: The actual age of an object (rock) or geologic event in years. Often determined by radioactive dating methods. See **relative age.**

absolute humidity: The actual amount (weight) of water vapor in a given volume of air. Directly related to vapor pressure.

absolute zero: Theoretically, the lowest possible temperature. No heat energy can be extracted from matter at this temperature.

absorption: The interception or taking in of electromagnetic energy by a material. When electromagnetic energy is absorbed by a material it is usually changed into other forms of energy, often heat, which causes a temperature rise.

actual evapotranspiration (E_a): In reference to the water budget, the total amount of water lost through evaporation and transpiration from an area in a given period of time. Usually expressed as a depth (mm of water).

adiabatic temperature change: A change in temperature that occurs without heat being added or removed. The cooling of air as it rises and expands and the warming of air as it descends and compresses are adiabatic temperature changes.

aerobic bacteria: Bacteria that require the presence of free oxygen to carry on their life functions.

aerosol: Small solid or liquid particles suspended in a gas. Suspended solid or liquid water particles are the aerosols in fog and clouds.

air mass: A large body of air in the lower atmosphere (troposphere) with approximately uniform air temperature, pressure, and moisture at any given level.

air pressure: See **atmospheric pressure.**

altitude: (1) The vertical distance (elevation) between a point and sea level. (2) The angle of a celestial object above the horizon. Usually expressed in degrees; often measured with a sextant.

anaerobic bacteria: Bacteria that do not require the presence of free oxygen to carry on their life functions. They are associated with polluted water.

angle of insolation: The angle at which the sun's rays hit the earth's surface. The higher the sun is in the sky, the higher the angle of insolation.

anticyclone: A high-pressure mass of air within the troposphere in which air moves out from the center. They rotate clockwise in the Northern Hemisphere and counterclockwise in the Southern Hemisphere. Also called *HIGHS.*

aphelion: The point in a planet's orbit when it is farthest from the sun. For the earth, aphelion occurs about July 4th, when it is about 152,000,000 kilometers from the sun.

apparent diameter: The diameter a celestial object (such as the sun or a planet) appears to have, depending on its distance from an observer; not the actual diameter. It is usually expressed in angular units. The closer a celestial object is to the earth, the larger its apparent diameter.

apparent solar day: A day of varying length determined by the time it takes for the sun to arrive at its highest point in the sky on two consecutive days at the same location. It is always slightly longer than the time it takes for one complete rotation of the earth. Can be measured by a sundial. See also **mean solar day.**

apparent solar time: Time based on the apparent solar day. Also called *local* or *sundial time.*

arc: A curved line that is part of a circle. An arc is the shape of the path of most celestial objects, such as the sun, in their daily paths through the earth's sky.

arid climate: A dry climate where the precipitation (P) is less than the potential evapotranspiration (E_p) for a large part of the year. Thus there is a deficit (D) of moisture and a drought much of the time.

asthenosphere: The layer of the earth's mantle below the lithosphere. It is a plastic, partly solid, partly liquid layer that allows tectonic plate movement.

astronomical unit: The mean distance of the earth from the sun—about 150,000,000 kilometers. Commonly used as a unit of distance in astronomy.

atmosphere: The shell of gases surrounding the earth. It is divided into layers according to differences in chemical and physical properties.

atmospheric pressure: The weight of the overlying atmosphere pushing down on a given unit of area. It is affected by changes in temperature, moisture, and altitude. Also called *air pressure* or *barometric pressure.*

atmospheric variables: The changeable conditions in the atmosphere that cause weather. The variables include temperature, air pressure, wind, and moisture.

axis: An imaginary line through the earth from the north to the south geographic poles, about which the earth rotates.

banding: The layered arrangement of mineral crystals in some metamorphic rocks due to the alignment and separation of like minerals.

barometric pressure: See **atmospheric pressure.**

bedrock: The solid and largely unweathered portion of the crust and lithosphere in an area that is underneath the soil and other loose material. Also called the *local rock.*

benchmark: A permanent marker, usually metal, at a specific location, that indicates an exact elevation or altitude at the time of installation.

calorie: A unit of heat energy defined as the quantity of heat needed to raise the temperature of 1 gram of water 1°C.

capillarity: The process by which water is drawn into pores due to molecular attraction.

capillary migration: The upward movement of water, against gravity, in part of the soil, regolith, or bedrock due to capillarity. Also called *capillary action.*

capillary water: The water held in pores in the soil and the rocks in the zone of aeration as a result of the process of capillarity.

carbon-14: A radioactive isotope of carbon with a short half-life (5700 years). Used only to date recent (up to 50,000-year-old) remains of organic material.

celestial object: Any object outside the earth's atmosphere, including moons, comets, planets, stars, and galaxies.

cementation: The process by which solid sediments are "glued" together by precipitated minerals, forming a sedimentary rock.

change: The observation or inference that the characteristics of a portion of the environment have altered.

change in soil storage (ΔSt): In reference to the water budget, the amount of water added to or taken from an area's soil storage. A positive change in soil storage is called *recharge,* and a negative change is called *usage.*

chemical sedimentary rocks: Monomineralic crystalline sedimentary rocks that form by evaporation and/or precipitation of dissolved minerals. They include rock salt and rock gypsum.

chemical weathering: The process whereby chemicals, such as oxygen and water, alter rocks and other earth materials, resulting in new minerals (chemicals). Rusting is a common example.

classification: The grouping together of objects and/or events with similar observed properties to make them more meaningful for study.

clay: See **colloids.**

climate: The total view or concept of an area's weather over a long period of time. It includes not only averages but also extremes.

cloud: A visible mass of suspended liquid water droplets and/or ice crystals in the atmosphere.

cold front: A weather front in which the colder air mass is pushing the warmer air mass. Cold fronts are characterized by a steep slope, rapid changes in weather, and often thunderstorms.

colloids: Very small solid particles (less than 0.0004 cm in diameter) that remain suspended in water for long periods of time. They make muddy water muddy. Also called **clay.**

compression: The process by which deposited sediment is pressed down by overlying sediments, water, and/or earth movements, resulting in the formation of sedimentary or metamorphic rocks. Also called *compaction.*

compression waves: In earthquakes, the waves that cause particles through which the waves travel to vibrate in the direction the waves are moving. Compression waves are the fastest-moving of earthquake waves. Also called *primary,* or *P, waves.*

condensation: The change in phase from a gas to a liquid, such as when water vapor changes to liquid water droplets as clouds form.

condensation surface: A solid surface required for the condensation of water. Aerosols, such as dust, salt crystals, and ice, act as condensation surfaces in the formation of clouds. In dew and frost formation, the condensation surfaces are the earth's surface features.

conduction: The transfer of heat energy from atom to atom, in any state of matter, through contact when atoms collide. Conduction of energy occurs most readily in solids, especially metals.

conservation of energy: The concept that energy is neither created nor destroyed, but remains the same in total amount in a closed system.

contact metamorphic zone: A special type of *transition zone* between rock types caused by the baking or altering of older bedrock by contact with molten rock (lava or magma). Changes older rock into nonsedimentary metamorphic rock.

continental climate: The climate of inland areas not moderated by a large body of water; characterized by hot summers and cold winters, and thus having a large annual temperature range.

continental crust: The part of the earth's crust (upper lithosphere) that makes up the continental blocks. Compared to oceanic crust, it is thicker, lower in density, and granitic rather than basaltic in composition.

continental drift: The concept that the continents have been and are presently shifting their position on the earth's surface; really due to plate tectonics.

continental polar air mass (cP): Very common cold and dry air masses that invade the contiguous United States from Canada.

continental tropical air mass (cT): Relatively rare hot and dry air masses that form in the southwestern United States or Mexico, that may affect weather of the contiguous United States in the summer.

contour line: An isoline on contour, or topographic, maps that represents points of equal elevation on the earth's surface.

contour map: A model of the elevation field of the earth's surface using contour lines and other symbols. Also called a *topographic map.*

convection: The transfer of heat energy by circulatory movements in a fluid (liquid or gas) that results from differences in density within the fluid.

convection current: A circulatory motion in a fluid due to convection. Also called *convection cell.*

convergence: (1) the coming together of air currents at the earth's surface and at the top of the troposphere. (2) The direct collision of tectonic plates in the **plate tectonic theory.**

convergent plate boundary: The interface between two colliding tectonic plates. Often associated with mountain building, ocean trenches, and volcanic island arcs.

coordinate system: A grid or system of lines for determining the location of a point on a surface.

core: The innermost region of the earth, thought to be composed of iron and nickel. The outer part of the core is thought to be liquid and the inner part solid.

Coriolis effect: The deflection of all moving particles of matter at the earth's surface, which provides evidence for the earth's rotation; the deflection is to the right in the Northern Hemisphere and to the left in the Southern Hemisphere.

correlation: In geology, the process of showing that rocks or geologic events in different places are similar in relative or absolute age.

crust: The outermost portion of the earth's solid lithosphere. It is separated from the uppermost mantle by the Moho interface.

crystal: (1) The individual mineral constituent units of many rocks. (2) A solid with a definite internal structure because its constituent atoms are arranged in a characteristic, regular, repeating pattern.

crystalline: A term used to describe a rock that is composed of intergrown mineral crystals.

crystallization: A type of solidification in which molten rock (magma or lava) cools to form nonsedimentary igneous rocks composed of minerals that have a definite arrangement of their atoms. Also see **solidification.**

cyclic change: Orderly changes in the environment in which the events constantly repeat themselves with reference to time and space.

cyclone: A low-pressure portion of the troposphere that has air moving toward its center. Cyclones in the Northern Hemisphere rotate counterclockwise and in the Southern Hemisphere they rotate clockwise. Types of cyclones include hurricanes, tornadoes, and mid-latitude cyclones. Also called a *LOW.*

daily motion: The apparent, usually east-to-west movement of celestial objects in the sky at 15 degrees per hour actually caused by the earth's west-to-east rotation. Daily motion causes the sun, stars, etc. to appear to move in circular or arc-shaped paths in the sky.

deficit (D): The condition in the local water budget when actual evapotranspiration (E_a) does not equal potential evapotranspiration (E_p) because there is not enough water from precipitation (P) and soil storage (St). A long period of deficit is a drought.

density: The ratio of the mass of an object to its volume. $\text{density} = \dfrac{\text{mass}}{\text{volume}}$

denudation: The stripping off and removal (erosion) of the plants, soil, and regolith at the earth's surface. Denudation is often caused or increased by the careless activities of human beings.

deposition: The process by which sediments are released, dropped, or settled from erosional systems; it includes the precipitation of dissolved minerals from water in the formation of chemical sedimentary rocks. Also called *sedimentation.*

desert: A region with an arid climate where the average yearly precipitation (P) is much smaller than potential evapotranspiration (E_p).

dew point: The temperature at which the air is saturated with moisture and the relative humidity is 100%. At temperatures below the dew point, condensation of water vapor occurs.

direct rays: Rays of sunlight that strike the earth at an angle of 90°. Also called *vertical rays* or *perpendicular insolation.*

distorted structure: The curving and folding of the foliations in nonsedimentary metamorphic rocks due to heat and compression.

divergence: (1) The spreading out of air from rising or falling currents of air in the troposphere. (2) The type of plate movement in which lithospheric plates spread or rift from each other, according to the **plate tectonic theory.**

divergent plate boundary: The interface between two tectonic plates that are spreading apart at a mid-ocean ridge.

duration of insolation: The length of time insolation is received at a location in a day, or how long the sun is in the sky in a day.

dynamic equilibrium: A condition of changing balance between opposing processes, such as evaporation and condensation or erosion and deposition.

earthquake: A natural, rapid shaking of the lithosphere caused when rocks are displaced due to the release of energy; most often associated with fault movement, but also associated with other events, such as volcanic eruptions.

eccentricity: The degree of ovalness of an ellipse, or how far an ellipse is from being a circle. Eccentricity is computed using the formula

$$\text{eccentricity} = \frac{\text{distance between foci}}{\text{length of the major axis}}$$

electromagnetic energy: Energy that is emitted (radiated) in the form of transverse waves, into any part of the universe. Electromagnetic energy radiates from all objects not at a temperature of absolute zero. Examples include visible light, radio waves, and infrared energy.

electromagnetic spectrum: A model, such as a chart, that shows the full range of the types of electromagnetic energy, usually in order of wavelength.

ellipse: A closed curve around two fixed points, called foci, in which the sum of the distances between any point on the curve and the foci is a constant. All planetary orbits, including the earth's, are elliptical in shape.

energy: The ability to do work.

environmental equilibrium: The balance that exists among the natural parts of the environment even though all parts of the environment are constantly changing.

epicenter: The place on the earth's surface lying directly above the point at which an earthquake originates (the focus).

equator: The parallel on earth midway between the geographic north and south poles with a latitude of 0 degrees.

equinox: A time when the sun is directly overhead at noon at the equator, and there are 12 hours of daylight and 12 hours of darkness over the whole earth. The spring (vernal) equinox is about March 21st, and the fall (autumnal) equinox is about September 23rd.

erosion: The carrying away of soil and pieces of rock by wind, water, ice, etc. Erosion is the process by which sediments are obtained and transported. Erosion also refers to the wearing away and lowering of the earth's surface.

error: The amount of deviation or incorrectness in a measurement. See also **percent error.**

escarpment: A steep slope or cliff in layered (stratified) rocks. Formed from certain rock layers that are resistant to weathering and erosion.

evaporation: The change in phase from liquid to a gas, such as liquid water into water vapor (steam). Also called *vaporization.*

evapotranspiration: The combination of the processes of evaporation and transpiration.

event: The name used to describe the occurrence of a change in the environment.

extrusion: A body of nonsedimentary igneous rock formed by the cooling and solidification of liquid rock (lava) at the earth's surface.

extrusive igneous rock: Igneous rocks, such as basalt and rhyolite, that form through the cooling and solidification of liquid rock (lava) at the earth's surface. Also called *volcanic igneous rock.*

fault: A crack in a mass of rock with displacement, or movement, of rock along the crack.

field: Any region of space that has some measurable value of a given quantity at every point, such as the earth's magnetic field.

focus (plural **foci**): (1) In an ellipse, either of two fixed points located so that the sum of their distances to any point on the ellipse is a constant. The sun is at one of the two foci of the orbit of each of the planets. (2) The place where an earthquake actually originates.

folded strata: The bends in layered rock due to movement in the lithosphere; a type of deformed strata.

foliation: The layering of mineral crystals in metamorphic rocks.

formation: See **rock formation.**

fossil: Any evidence of former life, either direct or indirect.

Foucault pendulum: A freely swinging pendulum whose path appears to change in a predictable way, thus providing evidence for the earth's rotation.

frames of reference: Properties or characteristics by which something can be described, studied, or compared. Time and space are frames of reference for studying change.

frictional drag: Friction at the interface of the atmosphere and the earth's surface caused by wind, air-mass movement, the Coriolis effect, etc. Heat produced by frictional drag minimally heats the atmosphere.

front: The interface between two air masses of different characteristics.

fusion: The change of state from a solid to a liquid. Also called *melting.*

geocentric model: An early concept of celestial objects and their motions in which all celestial objects revolved around the earth, which was stationary and was the center of the universe.

geographic poles: The North and South poles of the earth, with a latitude of 90 degrees. The geographic poles are located at opposite ends of the earth's axis of rotation.

geologic time scale: A chronological model of the geologic history of the earth using divisions called eons, eras, periods, and epochs. See the *Earth Science Reference Tables* for details.

geosyncline: A large, shallow ocean basin near continental margins that slowly subsides under large quantities of sediment. It is thought that geosynclines are eventually uplifted, forming mountains and continental areas. See **passive margin basin.**

graded bedding: A layering of sediment or sedimentary rock that shows a gradual change in particle size, with the largest particles on the bottom and the smallest ones on top.

gradient: The rate of change from place to place within a field. Also called *slope.*

$$\text{gradient} = \frac{\text{amount of change in the field}}{\text{distance through which change occurs}}$$

gravitation: The attractive force that exists between any two objects in the universe. It is proportional to the product of the masses of the objects and inversely proportional to the square of the distance between their centers.

$$\text{force} \propto \frac{\text{mass}_1 \times \text{mass}_2}{(\text{distance between their centers})^2}$$

gravity: The force that pulls objects toward the center of the earth.

greenhouse effect: A process that warms the atmosphere and reduces heat loss by terrestrial radiation from the earth's surface. It results from the fact that the atmosphere transmits the short-wave radiation received from the sun, but absorbs and is heated by the long-wave radiation from the earth's surface.

ground water: The portion of the subsurface water found beneath the water table; the water in the zone of saturation. Also see **subsurface water.**

half-life: The time required for one half of the atoms in a given mass of a radio-active isotope to decay, or change, to a different isotope.

heat energy: Energy that is transferred from one body to another as a result of a difference in temperature between the two bodies; also called *thermal energy.*

heliocentric model: The modern concept of celestial objects and their motions, in which the rotating earth and other planets revolve around the sun.

HIGH: See **anticyclone.**

horizontal sorting: The sorting of particles of sediments into layers in which particle size and density decrease in one horizontal direction—the direction toward which the erosional system was moving.

humid climate: A moist or wet climate where precipitation (P) equals or is greater than potential evapotranspiration (E_p) on a yearly average.

hydrosphere: The liquid water that rests on much of the earth's surface. The oceans constitute most of the hydrosphere.

igneous rock: A nonsedimentary rock formed by the cooling and solidification of molten material (lava or magma) above or below the earth's surface.

index fossil: A fossil used in correlation and relative dating of rocks. A species used as an index fossil generally lived for only a short time and was distributed over a large geographic area.

inference: An interpretation (conclusion, theory, or explanation) of observation(s).

infiltration: The seeping and entering of liquid water from the earth's surface into the ground (bedrock and soil), where the water becomes subsurface water.

inner core: The innermost zone of the earth's core, which is thought to be composed of iron and nickel in a solid state.

insolation (INcoming SOLar radiATION): The part of the sun's radiation that is received by the earth.

instrument: A device made by people, that aids or extends the human senses beyond their normal limits in order to obtain more accurate observations.

intensity of insolation: The relative strength of the sun's radiations intersecting a specific area of the earth in a specific amount of time, such as calories per square meter per minute. The higher the angle of insolation, the greater the intensity of insolation and the sun's heating effect.

interface: The boundary zone between regions with different properties. Energy is usually exchanged across an interface.

intrusion: A mass of nonsedimentary igneous rock formed by the cooling and solidification of liquid rock (magma) below the earth's surface.

intrusive igneous rock: An igneous rock, such as granite or gabbro, that forms by the cooling and solidification of liquid rock beneath the earth's surface. Also called *plutonic igneous rock.*

isobar: An isoline used on weather and climatic maps to connect points of equal air pressure.

isoline: A line used on a model of a field, such as a map, which connects points of equal value of a field quantity. Examples of isolines are isotherms, isobars, and contour lines.

isotasy: A principle that states that the earth's crust is in a state of equilibrium and that any change in the mass of one part of the crust will be offset by a change in mass of another part to maintain the equilibrium. Isotasy is often used to explain why the continental crust floats higher than the denser oceanic crust and why mountains and other continental areas rise as they are eroded.

iso-surface: A surface in a model of a three-dimensional field in which all points on the surface have the same field value.

isotherm: An isoline used on weather and climatic maps to connect points of equal air temperature.

isotope: One of the varieties of an element. The various isotopes of an element all have the same atomic number and chemical properties. However, they differ in their atomic masses and physical properties. For example, carbon-12 and carbon-14 are isotopes of carbon.

joint: A crack in rocks along which there has been no relative movement or displacement, such as there is in a fault.

kinetic energy: The energy of movement of a mass or object. The greater the velocity or speed of an object, the greater the kinetic energy.

landscape: The characteristics of the earth's surface at the interface between the atmosphere and the hydrosphere and lithosphere. Also called *topography.*

landscape region: A portion of the earth's surface with landscape characteristics that distinguish it from other areas. Some of the distinguishing characteristics are rock structure, elevation, degree of slope, and stream patterns. Also called a *physiographic region.*

latent heat: Energy absorbed or released by a substance during a change of phase. This transfer of energy occurs without a change in temperature.

latitude: Angular distance north or south of the equator; usually expressed in degrees. Minimum latitude is at the equator (0°), and maximum is at the geographic poles (90°N or 90°S).

latitudinal climatic patterns: East-west belts, or zones, of climate types on the earth caused by latitudinal changes in climatic factors, such as temperature, moisture, winds, and ocean currents.

leveling forces: Forces that operate constantly at or near the earth's surface and that break down rocks, transport material from higher to lower elevations, and tend to level off and lower the land. Leveling forces include weathering, erosion, denudation, deposition, and subsidence. Also called *destructional forces.*

lithosphere: The outer, solid, rocky part of the earth as distinguished from the hydrosphere and atmosphere. The crust is the part of the lithosphere above the mantle.

local noon: Noon, determined by when the sun is at its highest position on a given day at a specific location; it is also called *sundial noon* or *apparent solar noon.*

local water budget: See **water budget.**

longitude: Angular distance east or west of the prime meridian; usually expressed in degrees. Minimum longitude (0°) is at the prime meridian, which runs through Greenwich, England, and maximum longitude is 180°E or W. The International Date Line, which runs through the Pacific Ocean, follows the 180° meridian along most of its length.

LOW: See **cyclone.**

mantle: Largely solid intermediate zone between the earth's crust and the outer part of the core.

marine climate: A coastal climate moderated by the effects of a large body of water (ocean, lake, etc.). Such areas have warmer winters and colder summers than areas of similar latitude not near a large body of water, thus they have a small annual temperature range.

maritime polar air mass (mP): Cool and humid air masses that invade the contiguous United States from the oceans to the northeast and the northwest.

maritime tropical air mass (mT): A very common warm and humid air mass that invades the contiguous United States from the oceans to the south, east, and west.

mass: The quantity of matter in an object. Unlike weight, mass is not affected by location.

mean solar day: The 24-hour day established for convenience in time-keeping; it was derived by averaging the apparent solar days in a year.

measurement: An observational process of obtaining more precise or accurate dimensions of properties of objects and events by making comparison with some standard of reference, often using instruments. An example would be using a ruler.

meridian: North-south trending lines on maps or globes of the earth that have constant longitude.

metamorphic rock: A nonsedimentary rock formed without melting from other rocks (igneous, sedimentary, or other metamorphic rocks) within the lithosphere in response to heat, pressure, or chemical action. Metamorphic rocks are often associated with mountain-building processes.

mid-ocean ridge: Huge chain of largely underwater mountain ranges in the oceans; associated with lithospheric plate divergence or rifting, ocean-floor spreading, earthquakes, and volcanoes.

mineral: A naturally occurring, crystalline, inorganic solid with physical and chemical properties that vary within certain specified limits.

model: Any way of representing (illustrating) the properties of an object, event, or system. Models include graphs, drawings, charts, mental pictures, numerical data, or scaled physical objects.

Moho: Short for *Mohorovicic discontinuity.* The interface, or boundary zone, between the crust and the mantle.

moisture: Water vapor in the atmosphere or subsurface water.

moisture capacity: A measure of the total amount of water vapor the air can hold at a particular temperature; the maximum specific humidity at a particular temperature.

monomineralic: A rock composed of just one mineral, such as rock gypsum or rock salt.

mountain: A landscape characterized by relatively high elevations, many changes in slope, and steep slopes. Internally, mountains are characterized by distorted rock structures, such as faults, folds, and tilted rocks. Often composed of much nonsedimentary rock. Old mountains may be low in elevation and have little slope.

nonsedimentary rock: Rocks that do not form directly from sediments. Igneous and metamorphic rocks are nonsedimentary rocks.

North Star: See Polaris.

oblate spheroid: A sphere that is slightly flattened at the top and bottom (the polar regions) and slightly bulging at the middle (the equatorial region); the shape of the earth.

observation: An interaction of one of the human senses, with or without the aid of instruments, to an aspect of the environment; a reaction that is not an inference.

occluded front: Formed when an advancing cold weather front pushes into a warm front, causing the warm air mass to be lifted off the earth's surface, forming mid-latitude cyclones (LOWs).

ocean-floor spreading: The principle that the oceanic lithosphere spreads outward (plate divergence or rifting) at mid-ocean ridges. Also called *sea-floor spreading.*

oceanic crust: The portion of the earth's crust that is usually below the oceans and not associated with the continental blocks. Oceanic crust is thinner and higher in density than continental crust and is basaltic rather than granitic in composition.

orbit: The path of an object revolving around another object, such as the path of the earth around the sun.

orbital speed: The speed of an orbiting body along its orbit at a given time. Also called *orbital velocity.*

organic: Refers to an earth material that is composed of and/or was formed by life forms. *Inorganic* means "not organic."

organic evolution: The theory that new species of organisms arise by gradual transitional changes from existing species.

original horizontality: A principle that states that sedimentary rocks and some extrusive igneous rocks are originally formed in horizontal layers. There are exceptions to this principle, but most sedimentary rock not found in horizontal layers is thought to have been deformed after formation by crustal change.

orographic effects: The effects that mountains have on climate.

outcrop: Exposed bedrock without a cover of soil or regolith.

outer core: The zone of the earth between the mantle and the inner core. It is thought to be a liquid because shear waves from an earthquake do not go through it. It is believed to be composed of iron and nickel.

parallel: East-west trending lines on maps and globes that have constant latitude.

passive margin basin: An ocean basin at a continental margin that receives sediment deposition but is *not* the site of a tectonic plate boundary. Similar to a **geosyncline.**

percent error or **percent deviation:** The numerical amount, expressed as a percent, that a measurement differs from a given, standard, or accepted value.

perihelion: The point in a planet's orbit when it is closest to the sun. Perihelion for the earth occurs about January 3, when it is about 147,000,000 kilometers from the sun.

period: (1) The amount of time it takes a planet to make one orbit, or revolution, around the sun. This amount of time is called the *year* for that planet. (2) In geology, a part of the geologic time scale smaller than an era.

permeability: The degree to which a porous material (such as rock or soil) will allow fluids, such as water, to pass through it.

permeability rate: The speed at which fluids, like water, can pass through a porous material. The speed at which water moves from above to below the earth's surface, becoming subsurface water, is a special type called *infiltration rate.*

perpendicular insolation: See **direct rays.**

phase: (1) One of the three main forms of matter—liquid, solid, or gas. Also called *state.* (2) The varying amount of the lighted portion of the moon, Venus, or Mercury visible from the earth.

phase change: The change of a substance from one phase, or state, to another.

physical weathering: The mechanical or physical alteration of rock and other earth materials at or near the earth's surface into smaller fragments (sediments) without a change in the mineral or chemical composition of the materials. Frost action is the most common type.

plain: A landscape of low elevation and gentle slopes; usually characterized by horizontal rock structure, unless it is a remnant of an old mountain region.

planetary wind belts: East-west zones on the earth where the wind blows from one direction much of the time. An example is the prevailing southwest winds that blow over much of the United States.

plateau: A landscape of relatively high elevation with generally undistorted horizontal sedimentary rocks or extrusive igneous lava flows.

plates: The sections of the lithosphere that move around the earth's solid surface. Also called *lithospheric plates* and *tectonic plates.*

plate tectonic theory: A theory stating that the earth's lithosphere is divided into about 20 sections called plates. The plates can move up and down or sideways on a plastic part of the upper mantle called the asthenosphere. Plates diverging, converging, and sliding by each other result in many of the earth's physical features and events, including continent and mountain formation, volcanoes, and earthquakes; includes the concepts of continental drift and ocean-floor spreading.

Polaris: The star that is almost directly over the geographic North Pole of the earth. Also called the *North Star.*

pollutants: Substances or forms of energy that pollute the environment; they include solids, liquids, gases, life forms, heat, sound, and nuclear radiation.

pollution: The occurrence in the environment of a substance or form of energy in concentrations large enough to have an adverse effect on people, their property, or plant and animal life.

polymineralic: Refers to rocks that contain more than one mineral.

porosity: Amount of open space (pores) in rocks or soils compared to total volume.

potential energy: The energy possessed by an object as a result of its position or location, chemical conditions, or phase of matter.

potential evapotranspiration (E_p)**:** The amount of water that would be lost from an area through evaporation and transpiration over a given time if the water were available. Potential evapotranspiration is determined by the amount of heat energy available and the amount of surface area for evapotranspiration.

precipitation: (1) The falling of liquid or solid water from clouds toward the earth's surface. (2) A type of deposition in which dissolved substances come out of solution to form solids, as in formation of chemical sedimentary rocks.

present weather: The conditions or state of atmosphere for a short period of time at a location determined by comparison with a standard list produced by the U.S. Weather Service. A partial abbreviated list is found on the sample station model on page 95.

pressure gradient: The amount of difference in air pressure over a specific distance; the greater the pressure gradient, the greater the speed of the wind.

primary waves: See **compression waves.**

probability: The odds of some environmental change, such as rain or an earthquake, taking place.

P-waves: See **compression waves.**

radiation: (1) The emission or giving off of energy in the form of electromagnetic energy. (2) The method by which electromagnetic energy moves from place to place by way of transverse waves.

radiative balance: A condition in which an object gives off as much energy as it receives. Under such conditions its average temperature remains the same.

radioactive dating: The use of radioactive isotopes to determine the absolute age of rocks on geologic events.

radioactive decay: The natural spontaneous breakdown of the nucleus of unstable atoms into more stable atoms of the same or other elements. The process releases energy, continues at a constant rate for any particular radioactive isotope, and is not affected by changes in temperature, pressure, or other environmental conditions. Also called **radioactivity.**

recharge $(+ \Delta St)$**:** In reference to the water budget, the addition of water to the soil storage by infiltration. Recharge can occur only if the soil is not saturated and if precipitation (P) is greater than potential evapotranspiration (E_p).

recrystallization: A process in the formation of many metamorphic rocks by which some mineral crystals grow in size at the expense of other crystals or sediments without true melting.

reflection: A change in direction of waves when the waves strike the surface of a material, in which the waves leave the surface at the same angle at which they arrived.

refraction: A change in direction and velocity of waves when they pass from one medium into another with a different density.

regolith: All the unconsolidated material at the earth's surface, including the soil.

relative age: The age of rocks or geologic events as compared to other rocks or events with no reference to a specific year or absolute date.

relative humidity: The ratio of the actual amount of water vapor in the air to the maximum amount of water vapor the air can hold. It is often expressed as a percent. Relative humidity can be calculated by the following formula:

$$\text{relative humidity (\%)} = \frac{\text{specific humidity}}{\text{maximum specific humidity (capacity)}} \times 100$$

residual sediment: Weathered material that has remained in its place of origin.

residual soil: Soil formed from rocks and/or regolith under the soil; soil that has not been transported from its place of origin.

reversal of earth's magnetic polarity: The fact that the earth's magnetic field and poles switch polarity (north for south and south for north) in intervals of thousands of years but in no known cycle.

revolution: Movement of one body about another in a path called an *orbit.*

rifting: The process by which tectonic plates separate, spread, or diverge. Mid-ocean ridges and continental rift valleys are the result of rifting.

rock: Any naturally formed solid that is part of the earth's lithosphere. Most rocks are composed of one or more minerals; a few rocks, such as bituminous coal, obsidian, and coral, are not composed of minerals at all.

rock cycle: A model of the interrelationships of the different rock types, the materials they form from, and the processes that produce them.

rock formation: The basic unit of geologic mapping, consisting of a body of rocks with similar features.

rock-forming mineral: Any one of a small number of minerals (20–30) that are commonly found in rocks; most of them are silicates.

rock resistance: The ability of a body of rock to withstand erosion and weathering.

rock structure: The features of rock that can be observed in an outcrop. Structural features include folds, faults, joints, tilting, and thickness of strata.

rotation: The spinning of an object on its own axis, like a top.

runoff: All natural flowing of water at the earth's surface, including stream flow.

saturation: The condition of being filled to capacity. When the atmosphere contains all the water vapor it can hold, it is filled to saturation and the relative humidity is 100%.

saturation vapor pressure: The vapor pressure of a parcel of air when it is filled or saturated with water vapor; the vapor pressure when relative humidity is 100%.

scalar field: A field that can be totally described in terms of magnitude (amount) alone. Temperature and relative humidity are scalar fields.

scattering: The refraction and/or reflection of waves in various directions.

seasons: The divisions of the year with characteristic weather changes.

secondary wave: The type of earthquake wave that causes particles through which it travels to vibrate at right angles to the direction of the wave motion. Secondary waves will travel only through solids, not through liquids or gases. Also called *shear waves* or *S-waves.*

sediment: Particles or materials formed by the weathering or erosion of rocks or organic materials; particles or materials transported by erosional systems.

sedimentary rock: The rocks that form directly from sediments by processes of cementation, precipitation of minerals, dewatering, and compression.

seismic waves: The energy waves generated by an earthquake, including primary and secondary waves. Also called *earthquake waves.*

seismograph: An instrument used to record seismic waves.

senses: The five abilities or faculties—sight, touch, hearing, taste, and smell—by which one observes.

shear waves: See **secondary waves.**

silicon-oxygen tetrahedron: The most common constituent unit of minerals. This four-sided pyramid unit contains four atoms of oxygen and one atom of silicon.

sink: In an energy system, a region that has a lower energy concentration than its surroundings. Energy flows toward a sink.

soil: The part of the regolith that will support rooted plants. Soil is produced by weathering and the action of bacteria, plants, and animals.

soil association: A basic mapping unit or unit of soil classification composed of soils with similar characteristics, such as composition, structure, porosity, permeability, and fertility.

soil horizon: A vertical layer of soil with certain characteristics such as the high organic content of the top soil horizon. The combination of all the horizons in an area is called the area's *soil profile.*

soil storage (St): In terms of the water budget, the amount of liquid water stored in the soil.

solar noon: See **local noon.**

solar system: Our star, the sun, and the portion of the Milky Way Galaxy occupied by the objects that revolve around the sun, including planets, comets, and asteroids.

solidification: The processes by which a liquid changes phase to a solid, such as when molten rock (lava and magma) changes into igneous rocks. In the case of the igneous rock obsidian, there are no minerals, thus no orderly arrangement of atoms, because the lava solidified so rapidly. Most igneous rocks are produced by a type of solidification called crystallization. Also see **crystallization.**

solstices: The two times of the year when the vertical rays of the sun fall the farthest from the equator. At the summer solstice (about June 21st) the vertical rays fall on 23½° north latitude and the duration and angle of insolation are greatest for most of the Northern Hemisphere and least in the Southern Hemisphere. At the winter solstice (about December 21st) the vertical rays fall on 23½° south latitude and the duration and angle of insolation are greatest for most of the Southern Hemisphere and least in the Northern Hemisphere.

sorted particles: See **sorting of sediments**

sorting of sediments: The degree of similarity of size in the particles in a mass of sediments or sedimentary rocks. The greater the degree of similarity of particles the more **sorted** the sediments; the greater the difference in the size of particles the more **unsorted** the material.

source: In an energy system, a region that has a higher energy concentration than its surroundings. Energy flows from the source.

source region: The area of the earth's surface over which an air mass forms and acquires its characteristics.

species: The basic unit in the classification of life forms. All members of the same species are similar in body features, environment, and life habits. Our species is *sapien.*

specific heat: The amount of heat, in calories, needed to raise the temperature of 1 gram of a substance 1 degree Celsius; or the degree of difficulty a material offers to heating up or cooling off. Liquid water has the highest specific heat (1 cal/g/degree Celsius) of all common substances.

stages of landscape development: The stages (including Youth, Maturity, Old Age, and Rejuvenation) in the evolution of a landscape feature or region. The stage is characterized by certain features, including the types of dominant forces, the amount of slope, elevation, and the amount of change in slope.

stationary front: A weather condition in which the boundary between two air masses remains in the same position without moving.

strata: The layers, or beds, of sedimentary rock; *stratum* is the singular form.

stream bed: The bottom, or floor, of a stream.

stream discharge: The volume of water passing a certain spot in a stream in a given amount of time.

stream drainage pattern: An aerial view of the stream courses in an area. The patterns are determined by the shape of the landscape, rock structure, and rock resistance.

structure: See **rock structure.**

subduction: The tectonic process in which one of the plates at a convergent boundary sinks under the other plate and eventually melts into the asthenosphere. Ocean trenches and volcanic island arcs are results of subduction.

sublimation: The phase change from a solid directly to a gas or from gas to a solid with no intermediate liquid phase. Frost forms by sublimation.

subsidence: The sinking or depression of a part of the earth's surface.

subsurface water: All water found in the soil, regolith, and bedrock beneath the earth's surface. Also see **ground water.**

sundial: An ancient time-keeping device that uses the position of the sun to determine apparent solar, local, or sundial time.

superposition: A principle applied in the relative dating of layered sedimentary and some extrusive igneous rocks. It states that the youngest rock layer is found on top and that rock age increases with depth. There are many instances where this theory does not apply, such as in deformed rocks, and where there are igneous intrusions.

surplus (S): In terms of the water budget, liquid water that is not removed from the surface by infiltration or evapotransporation and thus becomes runoff. There is always a surplus when the soil is saturated (maximum St) and when precipitation (P) is greater than potential evapotranspiration (E_p).

S-waves: See **secondary waves.**

technology: The use of scientific information to serve human needs; the means by which a society provides objects required for human subsistence and comfort.

temperature: A measure of the average kinetic energy of the particles of a body of matter. Heat energy always flows from a higher temperature to a lower temperature.

terrestrial: (1) Relating to the earth as compared to any other part of the universe. (2) Relating to land or continental environments as compared to air or water environments.

terrestrial motions: The motions of the whole earth, including rotation and revolution about the sun.

terrestrial radiation: The electromagnetic energy given off by the earth's surface; it is mostly long wavelength infrared energy.

texture: Size, shape, and arrangement of mineral crystals or sediments in a rock.

thermal energy: See **heat energy.**

tilted strata: A type of deformed rock in which the strata, or layers, have been forced out of a horizontal position, usually by crustal movement.

topographic maps: See **contour maps.**

topography: See **landscape.**

track: The path of movement of an air mass and/or front. Tracks are often predictable, which helps in weather forecasting.

transformation of energy: The changing of energy from one form to another.

transform plate boundary: An interface at which plates slide sideways past each other. Also called a *lateral fault plate boundary.*

transition zone: An area where a mass of rock changes from one class (sedimentary, metamorphic, or igneous) to another. In many cases in a transition zone it is impossible to determine which class a particular segment belongs to. Also see **contact metamorphic zone.**

transpiration: A process by which plants release water vapor into the atmosphere as part of their life functions.

transported sediment: Weathered or eroded rock and organic materials that have been moved by an erosional system from their place of origin.

transported soil: Soil that has been moved from its place of original formation. Examples would be glacial soils common in the northern United States and the alluvial soils found around rivers and deltas.

transporting system: A system that accomplishes erosion. It includes an agent of erosion, a driving force, and the material that is transported.

transverse wave: A wave that vibrates at right angles to its direction of motion. Examples are electromagnetic radiations and secondary earthquake waves.

ultraviolet radiation: A form of electromagnetic energy of shorter wavelength than visible light. Most of the ultraviolet energy insolation is absorbed by gases of the atmosphere, such as ozone, before reaching the earth's surface.

unconformity: A break, or gap, in the rock record caused by the burial of an erosion surface by newer rocks or sediments. The rocks below an unconformity are older than those above it, unless there has been overturning of strata.

uniformitarianism: A principle stating that the geologic processes taking place today also took place in the past, and that we can interpret past events by studying present geologic processes. Uniformitarianism does not mean that geologic processes always occur at the same rate. Also, it does allow for the possibility that there were processes that occurred in the past (such as the earth's formation) that are not happening today.

unsorted particles: See **sorting of sediments.**

uplifting forces: Forces that originate beneath or within the earth's lithosphere that raise the land, build mountains, and cause continental growth. The uplifting forces include volcanic action, isotasy, earthquakes, continental drift, ocean-floor spreading, and plate tectonics. Also called *constructional forces.*

uranium-238: A radioactive isotope of uranium that decays to lead-206 with a half-life of 4.5 million years. Because of its long half-life, U-238 is useful for dating very old rocks.

usage ($-\Delta ST$): In terms of the water budget, the loss of water from soil storage by evapotranspiration. Usage occurs when potential evapotranspiration (E_p) is greater than precipitation (P) and soil storage (St) is greater than zero.

vaporization: See **evaporation.**

vapor pressure: The pressure exerted by water vapor in a given volume of air. It is a measure of the amount of water vapor in the atmosphere and is directly related to the absolute humidity of the atmosphere.

vector field: A field that must be described in terms of both magnitude and direction. Gravitational, magnetic, and wind fields are examples of vector fields.

vein: A sheetlike mineral deposit formed from a solution that has filled a crack or permeable zone in previously formed rocks; thus a vein is younger than the rocks into which it intrudes.

vertical rays: See **direct rays.**

visibility: The farthest distance that one can see a prominent object at the horizon with the naked eye. Fog, air pollution, and precipitation are common causes of low visibility.

volcanic ash: Small pieces of extrusive igneous rock shot into the air during a volcanic eruption. Volcanic ash is very important in geologic dating because all the ash from one eruption will be about the same age, no matter what type of rock it is found associated with; thus it is a time marker useful in correlation.

volume: The amount of space an object occupies.

walking the outcrop: A method of correlation done by actually following the continuity of the individual layers of formations in outcrops of bedrock. Generally, this is useful only for short distances, except in arid regions where a large expanse of bedrock may be uncovered.

warm front: A weather front in which a warmer air mass is pushing a colder air mass. Warm fronts are characterized by a gentle slope and long periods of precipitation.

water budget: A numerical model of an area's water supply. Often shown on a monthly basis using the average figures from data gathered over many years.

water cycle: A model, often in diagram form, used to illustrate the movement and phase changes of water at and near the earth's surface.

water table: The interface between the zone of saturation and the zone of aeration; the top of the zone below which the regolith and/or bedrock is saturated with liquid water.

water vapor: Water in the form of a gas.

wavelength: The distance between a point on a wave and the corresponding point on the next wave, such as the distance between two successive peaks in an electromagnetic wave.

weather: The condition of the atmospheric variables, such as temperature, pressure, wind, and moisture, at a location for a relatively short period of time.

weathering: The chemical and physical alterations of rock and other earth materials at or near the earth's surface. Weathering occurs through the action of water, chemical agents, and living things.

wind: The horizontal movement of air over the earth's surface.

year: The time it takes for a planet to make one revolution around the sun. The earth's year is 365¼ days. Also see **period.**

zone of aeration: The soil, regolith, or bedrock from the earth's surface down to the water table, where the pores are only partly filled with subsurface (capillary) water; air fills the rest of the pores.

zone of saturation: The portion of the earth near the earth's surface that is below the water table, where the pores in the soil, regolith, or bedrock are filled with subsurface water (ground water).

EARTH SCIENCE
REFERENCE TABLES

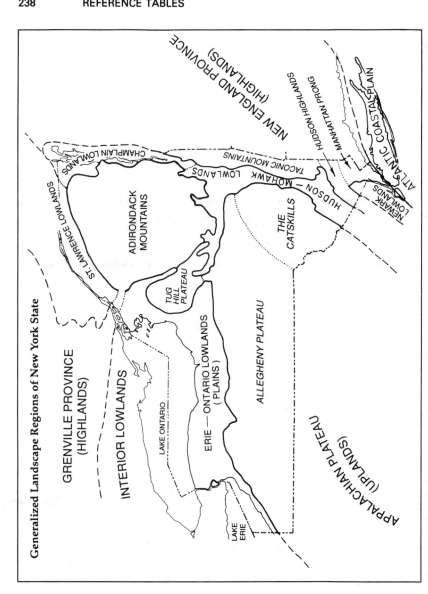

Generalized Landscape Regions of New York State

Surface Ocean Currents

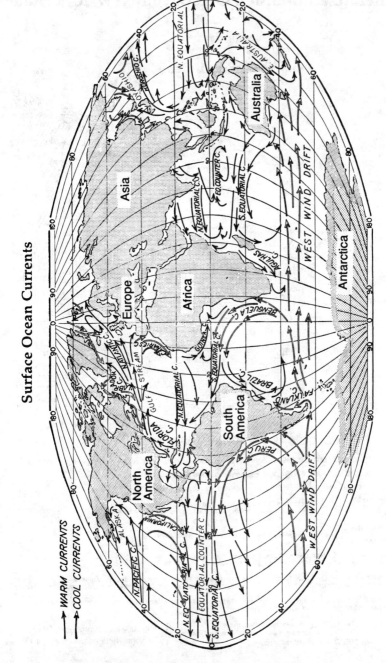

→ WARM CURRENTS
⇒ COOL CURRENTS

Generalized Bedrock Geology of New York State

COMPILED BY

GEOLOGICAL SURVEY

NEW YORK STATE MUSEUM

1989

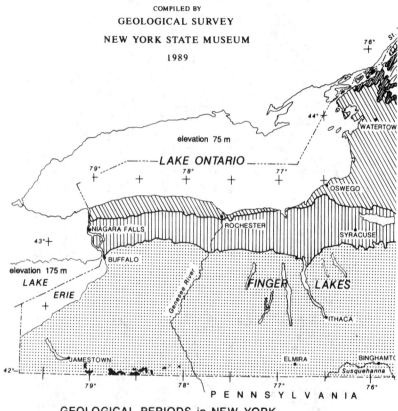

GEOLOGICAL PERIODS in NEW YORK

CRETACEOUS, TERTIARY, PLEISTOCENE (Epoch) unconsolidated gravels, sands, clays (not bedrock)

LATE TRIASSIC AND EARLY JURASSIC conglomerates, red sandstones, red shales, diabase

PENNSYLVANIAN and MISSISSIPPIAN conglomerates, sandstones, shales

DEVONIAN ⎱ limestones, shales, sandstones, conglomerates
SILURIAN ⎰ *Silurian also contains salt, gypsum and hematite.*

ORDOVICIAN ⎱ limestones, shales, sandstones, dolostones
CAMBRIAN   ⎰

CAMBRIAN and EARLY ORDOVICIAN sandstones, dolostones
Moderately to intensely metamorphosed east of the Hudson River.

CAMBRIAN & ORDOVICIAN (undifferentiated) quartzites, dolostones, marbles, schists
Intensely metamorphosed; includes portions of the Taconic Sequence and Cortlandt Complex.

TACONIC SEQUENCE sandstones, shales, slates. *Slightly to intensely metamorphosed rocks of*
CAMBRIAN and EARLY ORDOVICIAN *ages.*

MIDDLE PROTEROZOIC gneisses, quartzites, marbles
Lines are generalized structure trends.   Intensely Metamorphosed Rocks

MIDDLE PROTEROZOIC anorthositic rocks   (regional metamorphism about 1,000

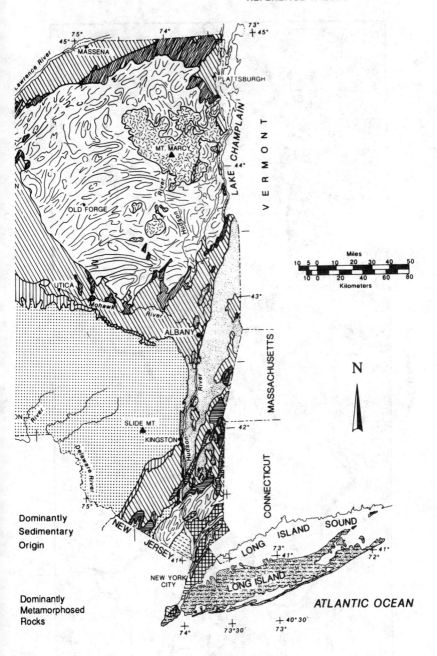

Dominantly
Sedimentary
Origin

Dominantly
Metamorphosed
Rocks

m.y.a.)

Tectonic Plates

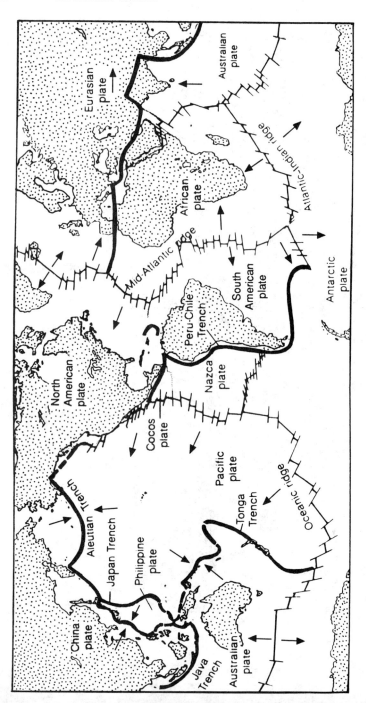

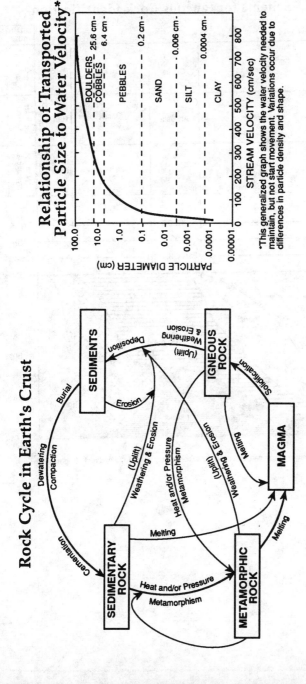

Relationship of Transported Particle Size to Water Velocity*

BOULDERS	25.6 cm
COBBLES	6.4 cm
PEBBLES	0.2 cm
SAND	0.006 cm
SILT	0.0004 cm
CLAY	

STREAM VELOCITY (cm/sec)

PARTICLE DIAMETER (cm)

*This generalized graph shows the water velocity needed to maintain, but not start movement. Variations occur due to differences in particle density and shape.

Rock Cycle in Earth's Crust

SEDIMENTS

Weathering & Erosion

Deposition

(Uplift)

IGNEOUS ROCK

Solidification

Burial

Erosion

Dewatering

Compaction

(Uplift) Weathering & Erosion

Heat and/or Pressure Metamorphism

(Uplift) Weathering & Erosion

Melting

MAGMA

Cementation

Melting

SEDIMENTARY ROCK

Heat and/or Pressure Metamorphism

METAMORPHIC ROCK

Melting

Scheme for Igneous Rock Identification

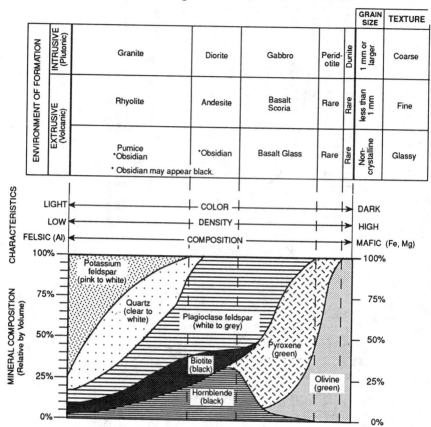

Note: The intrusive rocks can also occur as exceptionally coarse-grained rock, Pegmatite.

Scheme for Sedimentary Rock Identification

INORGANIC LAND-DERIVED SEDIMENTARY ROCKS

TEXTURE	GRAIN SIZE	COMPOSITION	COMMENTS	ROCK NAME	MAP SYMBOL
Clastic (fragmental)	Mixed, silt to boulders (larger than 0.001 cm)	Mostly quartz, feldspar, and clay minerals; May contain fragments of other rocks and minerals	Rounded fragments	Conglomerate	
			Angular fragments	Breccia	
	Sand (0.006 to 0.2 cm)		Fine to coarse	Sandstone	
	Silt (0.0004 to 0.006 cm)		Very fine grain	Siltstone	
	Clay (less than 0.0006 cm)		Compact; may split easily	Shale	

CHEMICALLY AND/OR ORGANICALLY FORMED SEDIMENTARY ROCKS

TEXTURE	GRAIN SIZE	COMPOSITION	COMMENTS	ROCK NAME	MAP SYMBOL
Nonclastic	Coarse to fine	Calcite	Crystals from chemical precipitates and evaporites	Chemical Limestone	
	Varied	Halite		Rock Salt	
	Varied	Gypsum		Rock Gypsum	
	Varied	Dolomite		Dolostone	
	Microscopic to coarse	Calcite	Cemented shells, shell fragments, and skeletal remains	Fossil Limestone	
	Varied	Carbon	Black and nonporous	Bituminous Coal	

Scheme for Metamorphic Rock Identification

TEXTURE		GRAIN SIZE	COMPOSITION	TYPE OF METAMORPHISM	COMMENTS	ROCK NAME	MAP SYMBOL
FOLIATED	Slaty	Fine	CHLORITE / MICA / QUARTZ / FELDSPAR / AMPHIBOLE / GARNET / PYROXENE	Regional	Low-grade metamorphism of shale	Slate	
	Schistose	Medium to coarse		Regional	Medium-grade metamorphism; Mica crystals visible from metamorphism of feldspars and clay minerals	Schist	
	Gneissic	Coarse		(Heat and pressure increase with depth, folding, and faulting)	High-grade metamorphism; Mica has changed to feldspar	Gneiss	
NONFOLIATED		Fine	Carbonaceous		Metamorphism of plant remains and bituminous coal	Anthracite Coal	
		Coarse	Depends on conglomerate composition		Pebbles may be distorted or stretched; Often breaks through pebbles	Meta-conglomerate	
		Fine to coarse	Quartz	Thermal (including contact) or Regional	Metamorphism of sandstone	Quartzite	
		Fine to coarse	Calcite, Dolomite		Metamorphism of limestone or dolostone	Marble	
		Fine	Quartz, Plagioclase	Contact	Metamorphism of various rocks by contact with magma or lava	Hornfels	

Inferred Properties of Earth's Interior

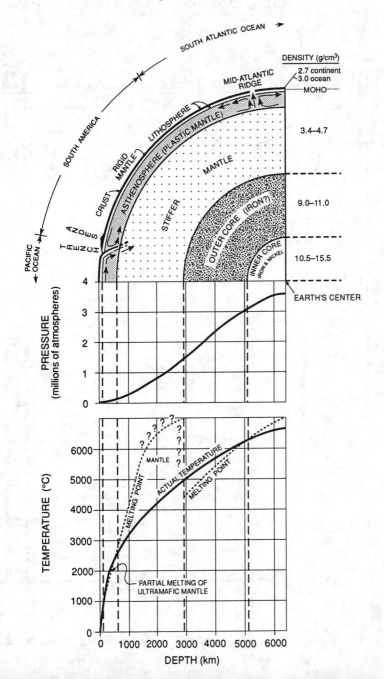

GEOLOGIC HISTORY OF NEW

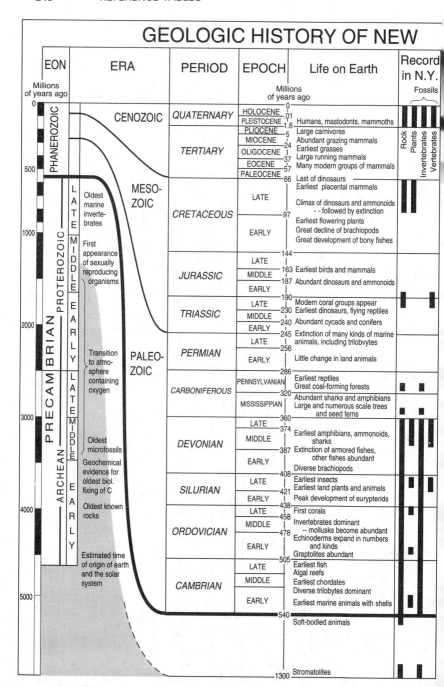

EON	ERA	PERIOD	EPOCH	Life on Earth	Record in N.Y.
Millions of years ago			Millions of years ago		Fossils

PHANEROZOIC

CENOZOIC

QUATERNARY — HOLOCENE .01 / PLEISTOCENE 1.6 — Humans, mastodonts, mammoths

TERTIARY —
PLIOCENE 5 — Large carnivores
MIOCENE 24 — Abundant grazing mammals / Earliest grasses
OLIGOCENE 37 — Large running mammals
EOCENE 57 — Many modern groups of mammals
PALEOCENE 66 — Last of dinosaurs / Earliest placental mammals

MESOZOIC

CRETACEOUS —
LATE 97 — Climax of dinosaurs and ammonoids - - followed by extinction
EARLY — Earliest flowering plants / Great decline of brachiopods / Great development of bony fishes

JURASSIC —
LATE 163 / MIDDLE 187 — Earliest birds and mammals
EARLY 190 — Abundant dinosaurs and ammonoids

TRIASSIC —
LATE 230 — Modern coral groups appear / Earliest dinosaurs, flying reptiles
MIDDLE 240 — Abundant cycads and conifers
EARLY 245 —

PALEOZOIC

PERMIAN —
LATE 256 — Extinction of many kinds of marine animals, including trilobvytes
EARLY 286 — Little change in land animals

CARBONIFEROUS —
PENNSYLVANIAN 320 — Earliest reptiles / Great coal-forming forests
MISSISSIPPIAN 360 — Abundant sharks and amphibians / Large and numerous scale trees and seed ferns

DEVONIAN —
LATE 374 / MIDDLE 387 — Earliest amphibians, ammonoids, sharks / Extinction of armored fishes, other fishes abundant
EARLY 408 — Diverse brachiopods

SILURIAN —
LATE 421 — Earliest insects / Earliest land plants and animals
EARLY 438 — Peak development of eurypterids

ORDOVICIAN —
LATE 458 — First corals
MIDDLE 478 — Invertebrates dominant -- mollusks become abundant
EARLY 505 — Echinoderms expand in numbers and kinds / Graptolites abundant

CAMBRIAN —
LATE — Earliest fish / Algal reefs
MIDDLE — Earliest chordates / Diverse trilobytes dominant
EARLY 540 — Earliest marine animals with shells

Soft-bodied animals

1300 — Stromatolites

PRECAMBRIAN — PROTEROZOIC / ARCHEAN

Oldest marine invertebrates

First appearance of sexually reproducing organisms

Transition to atmosphere containing oxygen

Oldest microfossils

Geochemical evidence for oldest biol. fixing of C

Oldest known rocks

Estimated time of origin of earth and the solar system

Rock / Plants / Invertebrates / Vertebrates

NEW YORK STATE AT A GLANCE

Important Fossils of New York	Tectonic Events Affecting Northeast N. America	Important Geologic Events in New York	Inferred Position of Earth's Landmasses
CONDOR, MASTODONT, FIG-LIKE LEAF		Advance and retreat of last continental ice	TERTIARY 59 million years ago
		Uplift of Adirondack region	
	Passive Margin / Rifting	Sandstones and shales underlying Long island and Staten island deposited on margin of Atlantic Ocean	CRETACEOUS 119 million years ago
		Development of passive continental margin	
		Kimberlite and lamprophere dikes	
COELOPHYSIS		Atlantic Ocean continues to widen	
		Initial opening of Atlantic Ocean	
		Intrusion of Palisades Still / Rifting	TRIASSIC 232 million years ago
		Massive erosion of Paleozoic rocks	
CLAM	Transform Collision	Appalachian (Alleghanian) Orogeny caused by collision of North America and Africa along transform margin	
AMMONOID, BRACHIOPOD, NAPLES TREE		Catskill Delta forms / Erosion of Acadian Mountains	PENNSYLVANIAN 306 million years ago
		Acadian Orogeny caused by collision of North America and Avalon and closing of remaing part of Iapetus Ocean	
PLACODERM FISH		Evaporite basins; salt and gypsum deposited	
EURYPTERID		Erosion of Taconic Mountains; Queenston Delta forms	DEVONIAN/MISSISSIPPIAN 363 million years ago
CORAL HEAD, GRAPTOLITE	Continental Collision / Subduction	Taconian Orogeny caused by closing of western part of Iapetus Ocean and collision between North America and volcanic island arc	
TRILOBITE		Iapetus passive margin forms	ORDOVICIAN 458 million years ago
	Rifting / Passive Margin	Rifting and initial opening of Iapetus Ocean	
STROMATOLITES		Erosion of Grenville Mountains / Grenville Orogeny: Ancestral Adirondack Mtns and Hudson Highlands formed / Subduction and volcanism / Sedimentation, volcanism	

Earthquake P-wave and S-wave Travel Time

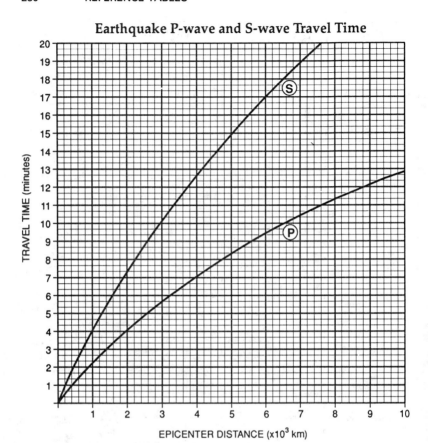

Average Chemical Composition
of Earth's Crust, Hydrosphere, and Troposphere

ELEMENT (symbol)	CRUST		HYDROSPHERE	TROPOSPHERE
	Percent by Mass	Percent by Volume	Percent by Volume	Percent by Volume
Oxygen (O)	46.40	94.04	33	21
Silicon (Si)	28.15	0.88		
Aluminum (Al)	8.23	0.48		
Iron (Fe)	5.63	0.49		
Calcium (Ca)	4.15	1.18		
Sodium (Na)	2.36	1.11		
Magnesium (Mg)	2.33	0.33		
Potassium (K)	2.09	1.42		
Nitrogen (N)				78
Hydrogen (H)			66	

Electromagnetic Spectrum

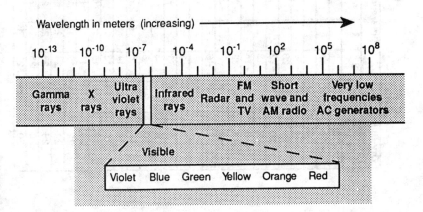

Dewpoint Temperatures

Dry-Bulb Temperature (°C)	Difference Between Wet-Bulb and Dry-Bulb Temperatures (C°)														
	1	2	3	4	5	6	7	8	9	10	11	12	13	14	15
-20	-33														
-18	-28														
-16	-24														
-14	-21	-36													
-12	-18	-28													
-10	-14	-22													
-8	-12	-18	-29												
-6	-10	-14	-22												
-4	-7	-12	-17	-29											
-2	-5	-8	-13	-20											
0	-3	-6	-9	-15	-24										
2	-1	-3	-6	-11	-17										
4	1	-1	-4	-7	-11	-19									
6	4	1	-1	-4	-7	-13	-21								
8	6	3	1	-2	-5	-9	-14								
10	8	6	4	1	-2	-5	-9	-14	-28						
12	10	8	6	4	1	-2	-5	-9	-16						
14	12	11	9	6	4	1	-2	-5	-10	-17					
16	14	13	11	9	7	4	1	-1	-6	-10	-17				
18	16	15	13	11	9	7	4	2	-2	-5	-10	-19			
20	19	17	15	14	12	10	7	4	2	-2	-5	-10	-19		
22	21	19	17	16	14	12	10	8	5	3	-1	-5	-10	-19	
24	23	21	20	18	16	14	12	10	8	6	2	-1	-5	-10	-18
26	25	23	22	20	18	17	15	13	11	9	6	3	0	-4	-9
28	27	25	24	22	21	19	17	16	14	11	9	7	4	1	-3
30	29	27	26	24	23	21	19	18	16	14	12	10	8	5	1

Relative Humidity (%)

Dry-Bulb Temperature (°C)	Difference Between Wet-Bulb and Dry-Bulb Temperatures (C°)														
	1	2	3	4	5	6	7	8	9	10	11	12	13	14	15
-20	28														
-18	40														
-16	48	0													
-14	55	11													
-12	61	23													
-10	66	33	0												
-8	71	41	13												
-6	73	48	20	0											
-4	77	54	32	11											
-2	79	58	37	20	1										
0	81	63	45	28	11										
2	83	67	51	36	20	6									
4	85	70	56	42	27	14									
6	86	72	59	46	35	22	10	0							
8	87	74	62	51	39	28	17	6							
10	88	76	65	54	43	33	24	13	4						
12	88	78	67	57	48	38	28	19	10	2					
14	89	79	69	60	50	41	33	25	16	8	1				
16	90	80	71	62	54	45	37	29	21	14	7	1			
18	91	81	72	64	56	48	40	33	26	19	12	6	0		
20	91	82	74	66	58	51	44	36	30	23	17	11	5	0	
22	92	83	75	68	60	53	46	40	33	27	21	15	10	4	0
24	92	84	76	69	62	55	49	42	36	30	25	20	14	9	4
26	92	85	77	70	64	57	51	45	39	34	28	23	18	13	9
28	93	86	78	71	65	59	53	47	42	36	31	26	21	17	12
30	93	86	79	72	66	61	55	49	44	39	34	29	25	20	16

Weather Map Information

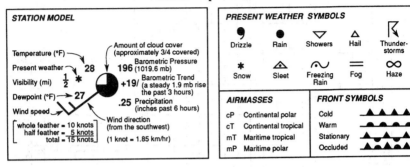

Lapse Rate

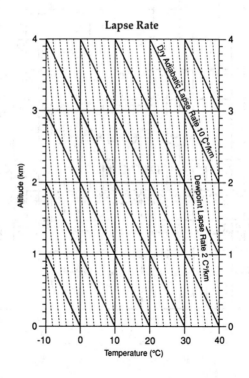

Temperature

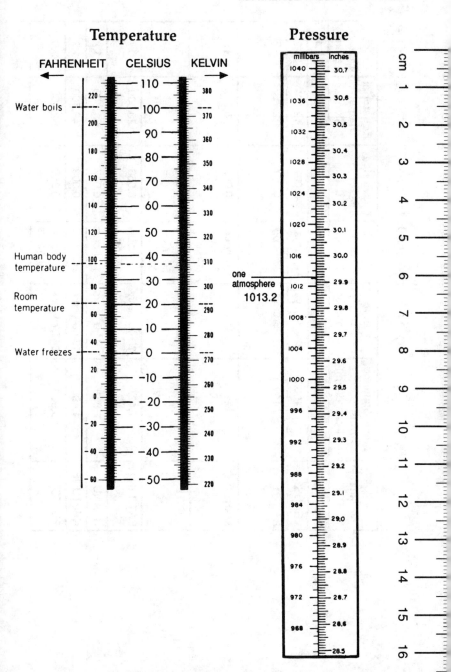

Pressure

Solar System Data

Planet	Mean Distance from Sun (millions of km)	Period of Revolution	Period of Rotation	Eccentricity of Orbit	Equatorial Diameter (km)	Density (g/cm³)
MERCURY	57.9	88 days	59 days	0.206	4,880	5.4
VENUS	108.2	224.7 days	243 days	0.007	12,104	5.2
EARTH	149.6	365.26 days	23 hours 56 min 4 sec	0.017	12,756	5.5
MARS	227.9	687 days	24 hours 37 min 23 sec	0.093	6,787	3.9
JUPITER	778.3	11.86 years	9 hours 50 min 30 sec	0.048	142,800	1.3
SATURN	1,427	29.46 years	10 hours 14 min	0.056	120,000	0.7
URANUS	2,869	84.0 years	11 hours	0.047	51,800	1.2
NEPTUNE	4,496	164.8 years	16 hours	0.009	49,500	1.7
PLUTO	5,900	247.7 years	6 days 9 hours	0.250	2,300	2.0

Selected Properties of Earth's Atmosphere

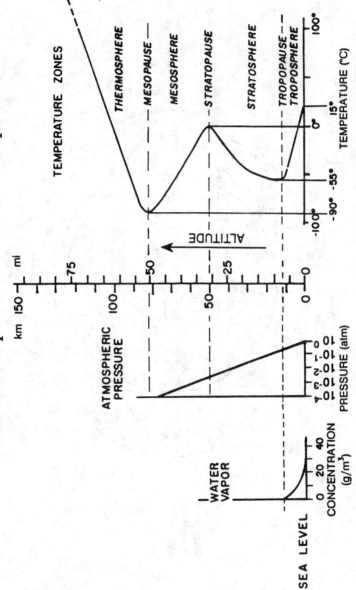

Planetary Wind and Moisture Belts in the Troposphere

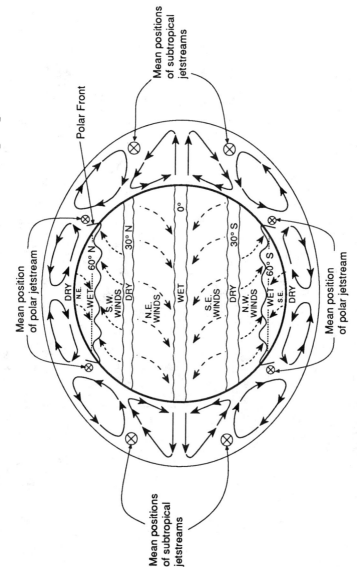

The drawing shows the locations of the belts near the time of an equinox. The locations shift somewhat with the changing latitude of the Sun's vertical ray. In the Northern Hemisphere the belts shift northward in summer and southward in winter.

Equations and Proportions

Equations

Percent deviation from accepted value	deviation (%) = $\dfrac{\text{difference from accepted value}}{\text{accepted value}} \times 100$
Eccentricity of an ellipse	eccentricity = $\dfrac{\text{distance between foci}}{\text{length of major axis}}$
Gradient	gradient = $\dfrac{\text{change in field value}}{\text{change in distance}}$
Rate of change	rate of change = $\dfrac{\text{change in field value}}{\text{change in time}}$
Circumference of a circle	$C = 2\pi r$
Eratosthenes' method to determine Earth's circumference	$\dfrac{\angle a}{360°} = \dfrac{s}{C}$
Volume of a rectangular solid	$V = \ell w h$
Density of a substance	$D = \dfrac{m}{V}$
Latent heat	$\begin{cases} \text{solid} \longleftrightarrow \text{liquid} \quad Q = mH_f \\ \text{liquid} \longleftrightarrow \text{gas} \quad Q = mH_v \end{cases}$
Heat energy lost or gained	$Q = m\,\Delta T C_p$

Proportions

Kepler's harmonic law of planetary motion	(period of revolution)$^2 \propto$ (mean radius of orbit)^3
Universal law of gravitation	force $\propto \dfrac{\text{mass}_1 \times \text{mass}_2}{(\text{distance between their centers})^2}$ $\left(F \propto \dfrac{m_1 m_2}{d^2} \right)$

C_p = specific heat
C = circumference
d = distance
D = density
F = force
h = height
H_f = heat of fusion
H_v = heat of vaporization
$\angle a$ = shadow angle
ℓ = length
s = distance on surface
m = mass
Q = amount of heat
r = radius
ΔT = change in temperature
V = volume
w = width
Note: $\pi \approx 3.14$

EURYPTERID
New York State Fossil

Physical Constants

Properties of Water

Latent heat of fusion (H_f) .... 80 cal/g

Latent heat of vaporization (H_v) .... 540 cal/g

Density (D) at 3.98°C .... 1.00 g/mL

Radioactive Decay Data

RADIOACTIVE ISOTOPE	DISINTEGRATION	HALF-LIFE (years)
Carbon-14	$C^{14} \rightarrow N^{14}$	5.7×10^3
Potassium-40	$K^{40} \nearrow Ar^{40} \searrow Ca^{40}$	1.3×10^9
Uranium-238	$U^{238} \rightarrow Pb^{206}$	4.5×10^9
Rubidium-87	$Rb^{87} \rightarrow Sr^{87}$	4.9×10^{10}

Astronomy Measurements

MEASUREMENT	EARTH	SUN	MOON
Mass (m)	5.98×10^{24} kg	1.99×10^{30} kg	7.35×10^{22} kg
Radius (r)	6.37×10^3 km	6.96×10^5 km	1.74×10^3 km
Average density (D)	5.52 g/cm³	1.42 g/cm³	3.34 g/cm³

Specific Heats of Common Materials

MATERIAL		SPECIFIC HEAT (C_p) (cal/g·C°)
Water	solid	0.5
	liquid	1.0
	gas	0.5
Dry air		0.24
Basalt		0.20
Granite		0.19
Iron		0.11
Copper		0.09
Lead		0.03

Index

Quick Review for Earth Science Regents Examination

This section provides a list of concepts and skills that have been tested most often on Earth Science regents exams. They are listed in order, from those tested most often to those tested least often. Each concept or skill listed is tested at least once each regents exam, on average. Where appropriate, the name of each concept or skill is followed by a list of the pages in this book where it is discussed. These page numbers are followed by the numbers of the corresponding questions at the end of the Topics. Finally, the related questions from recent regents exams are listed (copies of these exams are printed in the back of this book).

1. **Interpreting data tables and diagrams**
   1995: 6, 76–80, 86, 88
   1996: 76, 77, 85, 87, 88, 98, 100
   1997: 24, 36, 89
   1998: 53, 81, 86, 90

2. **Using data from the *Generalized Bedrock Geology of New York State* map in the *Earth Science Reference Tables***
   Questions: pp. 24–28: 18
   pp. 194–202: 1–3, 21, 22
   1995: 46, 53–55, 100
   1996: 50, 96, 97, 102
   1997: 51, 54, 77, 98–100
   1998: 20, 36, 54, 99, 100

3. **Using data from *Geologic History of New York State at a Glance* in the *Earth Science Reference Tables***
   Pages: 165–166, 188
   Questions: pp. 96–104: 51, 55–58, 60–67
   p. 105: 6–13
   1995: 42, 45, 46, 49, 91, 97
   1996: 46, 49, 51
   1997: 45, 51.86, 88, 90
   1998: 47, 50, 93, 95, 105

4. **Reading and interpreting isolines**
   Pages: 19, 20
   Questions: p. 22: 1, 6, 7, 10
   pp. 24–28: 23, 24
   1995: 72, 81–83
   1996: 57–60
   1997: 6, 22, 61–65
   1998: 61–63

5. **Using the *Weather Map Information* in the *Earth Science Reference Tables***
   Pages: 92, 93, 95
   Questions: pp. 96–104, 51, 55–58, 60–67
   p. l05: 6–13
   1995: 71, 73–75
   1996: 21, 73, 74
   1997: 71–75
   1998: 21, 71, 72

6. **Using latitude and longitude to locate places on the earth**
   Pages: 17–19
   Questions: pp. 22: 2, 5
   pp. 24–28: 2, 16–18
   p. 28: 7
   pp. 214–219: 5

   1995: 56–60, 105
   1996: 71, 102
   1997: 68, 69, 76–79, 100
   1998: 54

7. **Understanding how angle/intensity of insolation and duration of insolation change with the seasons on earth**
   Pages: 33–35, 67–71
   Questions: pp. 44–52: 17–22, 25
   pp. 75–81: 13–15, 17, 19–23
   1995: 10, 58
   1996: 10, 17, 67
   1997: 15, 67, 79, 80
   1998: 11, 16, 69, 70

8. **Using data from the *Generalized Landscape Regions of New York State* map in the *Earth Science Reference Tables***
   Page: inside back cover and two preceding pages
   Questions: pp. 214–219: 1, 5, 7
   1995: 55, 58, 100
   1996: 50, 99
   1997: 54, 96
   1998: 36, 54, 96, 99, 100

9. **Interpreting and extrapolating graphs**
   Questions: pp. 96–104: 37–39
   pp. 119–125: 18–22
   1995: 1, 76
   1996: 1, 72, 76
   1997: 91, 92
   1998: 24, 32

10. **Using data from the *Scheme for Sedimentary Rock Identification* in the *Earth Science Reference Tables***
    Pages: 147–148
    Questions: pp. 157–160: 3–5
    p. 160: 5
    1995: 35, 104
    1996: 91, 94
    1997: 32, 37, 87
    1998: 30, 33, 46, 56, 84

11. **Using data from the *Earthquake P-wave and S-wave Travel Time* graph in the *Earth Science Reference Tables***
    Pages: 170–171
    Questions: pp. 174–182: 32, 34, 35, 38–42
    1995: 87, 88, 90
    1996: 38, 39, 43
    1997: 41
    1998: 41, 88

12. Determining compass directions on maps
Page: 20
Questions: p. 22: 8, 9
           pp. 24–28: 24
           pp. 96–104: 64
1995: 11, 56
1996: 58, 100
1997: 23, 63, 97
1998: 20, 61

13. Understanding the math concepts of scientific notation, addition, subtraction, multiplication, division, and averaging
1995: 2, 77
1996: 48, 77, 98
1997: 1, 69, 76
1998: 103

14. Recognizing the locations and features of erosion and deposition by streams
Pages: 131, 141
Questions: pp. 142–145: 7, 8, 11, 13
           p. 145: 1, 4, 5, 7
1995: 85
1996: 30, 31
1997: 31, 85
1998: 28, 77, 78

15. Using data from the *Relationship of Transported Particle Size to Water Velocity* graph in the *Earth Science Reference Tables*
Pages: 132
Questions: pp. 133–137, 25, 30–31
1995: 25, 82, 83
1996: 32, 103
1997: 30, 104
1998: 56

16. Understanding the water budget concepts of surplus, deficit, recharge, usage, and soil moisture storage
Pages: 109–113
Questions: pp. 119–125: 18, 19, 21, 24, 25
           p. 126: 1
1995: 70, 78–80
1996: 28
1997: 27, 28
1998: 24

17. Using the concept of superposition to determine the relative age of rock layers
Pages: 183–184
Questions: pp. 194–202: 4–7, 12, 27
1995: 44, 50
1996: 93, 95
1997: 50, 89
1998: 45, 46

18. Recognizing glacial features of landscapes
Pages: 132, 140, 208
Questions: pp. 133–137: 33
           pp. 142–145: 4, 9
           pp. 214–219: 17, 18

1995: 51, 97, 99
1996: 33
1997: 26
1998: 31, 34, 48

19. Understanding the meaning and use of index fossils in rock dating
Page: 87
Questions: pp. 194–202: 9, 14, 15, 19
1995: 92, 95
1996: 47, 93, 95
1997: 48
1998: 52

20. Understanding the methods of mineral identification, including hardness, streak, color, cleavage, fracture, luster, and density
Pages: 154, 273, 274
Questions: pp. 157–160: 20, 21
1995: 34
1996: 86–89
1997: 34
1998: 35

21. Understanding how wind directions are determined by the earth's rotation and differences in air pressure
Pages: 84, 85, 92, 93
Questions: pp. 96–104: 11, 12, 17, 20
1995: 17, 30
1996: 22
1997: 11, 75
1998: 75

22. Understanding the relationships between resistance of rocks to weathering and erosion and the shape of landscape features
Page: 210
Questions: pp. 133–137: 5
           pp. 214–219: 19, 23
           pp. 219–220: 7
1995: 51, 97, 99
1996: 54
1997: 53
1998: 98

23. Understanding the relationships of the phase change of water to latent heat and temperature, the use of latent heat data, and the equation in the *Earth Science Reference Tables*
Pages: 57–58
Questions: pp. 60–64: 30, 32, 33, 36, 40
1995: 68
1996: 16
1997: 93–95
1998: 102

24. Understanding the relationships among temperature, dew point, and the chances of condensation (or sublimation) and precipitation
Page: 94
Questions: pp. 96–104: 4, 40, 42, 44

1995: none
1996: 42
1997: 42, 43
1998: 6

38. Using data from *Selected Properties of Earth's Atmosphere* in the *Earth Science Reference Tables*
Page: 16
Questions: pp. 24–28: 11, 12
           pp. 96–104: 1
1995: 9
1996: 6
1997: 102
1998: 6

39. Understanding eccentricity of an orbit and determining eccentricity using the *Equation and Solar System Data* table in the *Earth Science Reference Tables*
Page: 37
Questions: pp. 44–52, 33–35
1995: 61, 65
1996: 63
1997: none
1998: 101

40. Determining dewpoint temperature or percent relative humidity using the *Dewpoint Temperatures and Relative Humidity (%)* tables in the *Earth Science Reference Tables*
Pages: 89–90
Questions: pp. 96–104: 35
           p. 105: 2, 3
1995: 21
1996: 104
1997: 18
1998: 17

41. Understanding the oblate spheroid shape of the earth and the reasons we know the earth's shape
Pages: 12–14
Questions: pp. 24–28: 1, 3–6
1995: 7
1996: 7
1997: 4
1998: 4

42. Understanding the relationship between the altitude of Polaris (North Star) and latitude
Page: 13
Questions: pp. 24–28: 2, 5
1995: 60
1996: 102
1997: 70
1998: 3

43. Determining percent of deviation (error) using the formula in the *Earth Science Reference Tables*
Page: 2
Questions: pp. 4–7: 9, 15

1995: 102
1996: 3
1997: 3
1998: 37

44. Understanding the meaning of gravitation and the use of the proportion for Universal Law of Gravitation in the *Earth Science Reference Tables*
Page: 38
Questions: pp. 44–52: 38–40
1995: 64
1996: 62
1997: 9
1998: 9

45. Understanding how weather (storms, lows, etc.) moves from southwest to northeast over most of the United States
Pages: 86, 94, 116–117
Questions: pp. 96–104: 12, 13
           pp. 119–125: 35
1995: none
1996: none
1997: 23, 74
1998: 20, 49

46. Using data from *Scheme for Metamorphic Rock Identification* in the *Earth Science Reference Tables*
Pages: 149–150
Questions: pp. 157–160: 14
           p. 160: 4, 6, 7
1995: none
1996: 37
1997: 32
1998: 30, 83

47. Understanding Eratosthenes' method of determining Earth's circumference and using the formula in the *Earth Science Reference Tables*
Page: 15
Questions: pp. 24–27: 7
1995: 5
1996: 5
1997: 105
1998: 12

48. Understanding the relationships of surface properties (color and roughness) and the absorption and reflection of electromagnetic energy
Pages: 54, 66–67
Questions: pp. 60–64, 4, 5
1995: 16
1996: 68
1997: 16
1998: 67

Earth Science Performance Test

Ten points (out of 100) of the Earth Science Regents grade are derived from a performance test designed to evaluate various laboratory and classroom procedures refined during the yearly course of study. This performance test is normally given about two weeks before the rest of the regents exam. The test is divided into 6 groups of tasks called stations, with students given 6 minutes for each station. The 6 stations described here are those that have been used for the last few years. What will be used in the future may or may not be similar.

STATION 1: MINERAL PROPERTIES AND IDENTIFICATION

This station asks you first to determine four mineral characteristics (properties) for each of two samples. To aid in this process of mineral property determination, you are provided with a glass hardness plate, a porcelain streak plate, and a hand lens. You are asked to determine the following four mineral properties of each mineral sample. After determining the four properties of the two samples, you are asked to identify the samples with the aid of an identification table.

1. Cleavage.   Cleavage is a property of minerals that results in parallel smooth to semi-smooth surfaces (sides) when the mineral is broken. Samples are pre-broken for the performance test. You should look over the whole sample for parallel sets or a set of flat to semi-flat sides. Cleavage surfaces usually have a bright luster or shine, especially when compared with fracture surfaces (non-smooth breaking surfaces).

2. Streak.   Streak is the color of the powder of a mineral. It is used in identifying minerals because although the color of a mineral may vary, the color of the streak is consistent. To powder a mineral, rub the mineral sample across the porcelain streak plate (usually just once). The streak plate acts like a piece of sandpaper. Blow any excess powder off the streak plate and observe the color of the streak (line of powder) in a bright light. Often streak is first classified as white (no streak) or colored, and then further classified into specific colors.

3. Hardness.   Mineral hardness is not how easily a mineral breaks, as is often commonly believed, but how easily it can be scratched or dented. Mineral hardness is often compared to a standard reference set of ten minerals, the Moh's Hardness Scale. Another way of classifying hardness is to use window glass to divide minerals into soft and hard. To see if a mineral is soft or hard, hold the glass plate firmly on a desk and rub the mineral sample once or twice across the glass. A soft sample will not scratch the glass, but a hard sample will. To distinguish a scratch from a streak on the glass, rub the area of the possible scratch to see whether the streak comes off or the scratch remains. You can feel a true scratch with a fingernail or see the indentation with the hand lens.

4. Luster.   Luster is the way a mineral looks or shines in reflected light. To determine a mineral's luster, hold the sample up so that light reflects off the sample to your eyes. Luster is classified as either metallic or nonmetallic. A metallic luster is the shine of polished metal, like a new penny, brass doorknob, or chrome on a car. Nonmetallic lusters are further classified into types like glassy, dull, and pearly.

STATION 2: ROCK PROPERTIES AND CLASSIFICATION

At this station you are first asked to classify two different rock samples as sedimentary, igneous, or metamorphic. To aid in this classification you are provided with a hand lens and the three schemes for rock identification found in the reference tables.

Next you are asked to state a reason for your two classifications in two *complete* sentences. Be careful to start sentences with a capital letter, end them with a period, and include a verb and noun, or credit will be deducted.

The following are some suggested rock characteristics to aid the rock classification process and to include as facts in your supporting sentences. Not all characteristics will fit all samples.

Sedimentary rocks often have (1) bedding stratification or layering of sediments; (2) rounded grains, clasts, fragments, or sediments; (3) fossils; (4) cemented sediments with visible pores or openings; (5) fragments of other rocks.

Igneous rocks often have (1) polyminerallic composition; (2) crystalline texture; (3) interconnected mineral crystals with *no* layering of component parts; (4) glassy texture; (5) rounded gas pores or spaces.

Metamorphic rocks often have (1) polyminerallic composition; (2) interconnected mineral crystals *with* layering (foliation); (3) slaty, schistose, or gneissic foliation; (4) distorted or wavy rock structure; (5) stretched pebbles; (6) a high percent of mica minerals.

STATION 3: ANGULAR AND LENGTH MEASUREMENTS

At this station you are asked to make an angular measurement of a simulated sun's daily motion path drawn on a plastic hemisphere (sky model). You are provided with a 10 A.M. reference time on the sun's path and are given a later time of day in the regents directions. Using the fact that the sun "moves" at 15° per hour, you must calculate how many degrees the sun "moves" in the given time. As an example, suppose the given time is 5 P.M. The difference between 5 P.M. and 10 A.M. is seven hours. Seven hours times 15 degrees per hour equals 75 degrees. You are then asked to put a strip of masking tape over the sun's path on the dome. First you mark the 10 A.M. position on the tape, and then the position of the later time. The later time position is measured with the use of an external protractor (75 degrees in this example) and another mark is placed on the masking tape to show this later time position.

In making these angle measurements with the external protractor, take care to place the protractor directly over the sun's path. Be sure not to bend the protractor, as this will cause an erroneously low answer.

Next, you must remove the masking tape from the plastic hemisphere and place it on the answer sheet. Then you are asked to measure the distance between the two dots on the tape (the one for 10 A.M. and the one for the later time) using a metric ruler. The distance measurement should be recorded to the nearest 0.1 (tenth) of a centimeter. You are cautioned to be sure to include a number in the tenths position and not to include a hundredths number. Also be sure *not* to use the inches part of the ruler.

STATION 4: MASS AND VOLUME DETERMINATION

At this station, you are asked to use a balance to measure the mass of a mineral specimen to the nearest 0.1 (tenth) of a gram. You should be sure to include a number for the tenths place and use the 0.01 (hundredths) place to help determine the nearest tenth, but *don't record* a hundredths place.

Next you look up the name of the mineral specimen in the regents directions and find its density in a chart. Using the measured mass and the density from the chart, you are now asked to calculate the volume of the mineral using the formula

$$\text{density} = \frac{\text{mass}}{\text{volume}}, \text{ which can be converted to}$$

$$\text{volume} = \frac{\text{mass}}{\text{density}}.$$

To determine the volume, divide the mass by the density. Answers should be expressed to the nearest 0.1 (tenth) of a cubic centimeter (not to the hundredth or to the nearest whole number). As an example, suppose a sample of topaz has a density of 3.5 g/cm^3 and you calculate the mass to be 357.4 grams. Dividing the mass (357.4 grams) by the density (3.5 g/cm^3), you find that the volume equals 102.1 cm^3.

STATION 5: BEAD SETTLING TIME

At this station you are asked to use a stop watch to determine the time it takes for three different sizes of spherical beads to settle (drop) between two lines on a vertical column filled with a liquid. (You are provided with a calculator to aid with the math.) This is accomplished by dropping the beads and timing their settling. The time for each size bead should be measured twice and recorded to the nearest 0.1 (tenth) of a second. You should be sure not to confuse whole seconds, tenths of seconds, and hundredths of seconds on the stop watches. Also, if the two settling time readings for any one bead size are *not* similar (within 0.3 to 0.4 seconds of each other) then you should perform other settling measurements and pick values based on consistency. Caution: When starting and stopping the stop watch for the bead settling times, make sure your eyes are level with the lines on the columns as the beads pass by.

Next you are asked to average your two settling time readings for each of the three bead sizes and record these answers to the nearest 0.1 (tenth) of a second. To obtain the average, add the two readings together and then divide the result by two. Example: If a 5.0-millimeter bead settles in 8.7 seconds and 8.3 seconds, find the average by adding 8.7 seconds and 8.3 seconds (equals 17.0 seconds) and then dividing by 2 (equals 8.5 seconds).

STATION 6: LINE GRAPHING OF BEAD DATA

In station 6 you use the average bead settling times from station 5 and plot the data on a graph provided, which has scales, labels, and axis numbers. You need to be careful to plot (marking reasonably-sized dots) the data points to the nearest 0.1 (tenth) of a second and to the nearest 0.1 (tenth) of a millimeter. Use a pencil so you can make corrections clearly.

After the three bead settling times are plotted, connect the three points with a line, using an appropriate method (dot to dot or best fit). Be sure *not* to extend the line on the graph beyond the limits of lowest and highest timing values for the three bead sizes. Finally, you are asked to infer the settling time of a *given* imaginary bead size. You obtain this inference by reading the location on the line on your graph that corresponds to the diameter of the imaginary bead size.

DOING WELL ON THE PERFORMANCE TEST

There are approximately 32 tasks on the performance test, which are given a raw score of 0 to 25. This raw score is then converted to regents credits from 0 to 10 points. A raw score between 23 and 25 equals 10 regents credits; 21 to 22 raw points equal 9 regents credits.

How well you perform on this test is determined by the accuracy and precision of your answers. You are encouraged to carefully read the directions given with the test and to record your answers neatly and accurately. Be especially careful to answer the questions with the degree of accuracy asked for, not less and not more. As an example, if the directions ask for an answer to the nearest 0.1 (tenth) of a gram, don't just give a whole number or extend your answer to the nearest 0.01 (hundredth). Follow the directions. Another common problem is using or recording the wrong units, such as using inches instead of centimeters or millimeters instead of centimeters.

REGENTS EXAMINATIONS

EARTH SCIENCE

June 20, 1995

PART I

Answer all 55 questions in this part. [55]

Directions (1–55): For *each* statement or question, select the word or expression that, of those given, best completes the statement or answers the question.

1. The graph below shows the relationship between the mass and volume of a mineral.

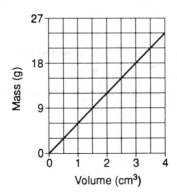

What is the density of this mineral?
(1) 6.0 g/cm³
(2) 9.0 g/cm³
(3) 3.0 g/cm³
(4) 4.5 g/cm³

2. A centimeter is 0.01 meter. This measurement can also be expressed as
(1) 1×10^{-1} m
(2) 1×10^{-2} m
(3) 1×10^{0} m
(4) 1×10^{2} m

3. Nine rock samples were classified into three groups as shown in the table below.

Group A	Group B	Group C
Granite	Shale	Marble
Rhyolite	Sandstone	Schist
Gabbro	Conglomerate	Gneiss

This classification system was most likely based on the
(1) age of the minerals in the rock
(2) size of the crystals in the rock
(3) way in which the rock formed
(4) color of the rock

4. An insulated cup contains 200 milliliters of water at 20°C. When 100 grams of ice is added to the water, heat energy will most likely flow from the
(1) water to the ice, and the temperature of the mixture will drop below 20°C
(2) water to the ice, and the temperature of the mixture will rise above 20°C
(3) ice to the water, and the temperature of the mixture will drop below 20°C
(4) ice to the water, and the temperature of the mixture will rise above 20°C

5. The diagram below shows the altitude of the noon Sun as measured on March 21 by observers at locations A and B.

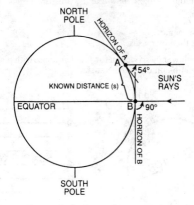

According to the *Earth Science Reference Tables,* an observer can use the known distance, s, and the Sun's altitude at A and B to find the Earth's
(1) density
(3) circumference
(2) eccentricity
(4) oblateness

6. A student drew the phase of the Moon observed from one location on the Earth on each of the dates shown below.

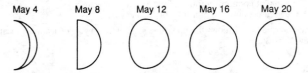

Which diagram best shows the Moon's phase on May 24?

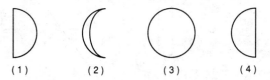

7. The best evidence that the Earth has a spherical shape would be provided by
(1) the prevailing wind direction at many locations on the Earth's surface
(2) the change in the time of sunrise and sunset at a single location during 1 year
(3) the time the Earth takes to rotate on its axis at different times of the year
(4) photographs of the Earth taken from space

8. A contour map shows two locations, X and Y, 5 kilometers apart. The elevation at location X is 800 meters and the elevation at location Y is 600 meters. What is the gradient between the two locations? [Refer to the *Earth Science Reference Tables.*]
(1) 12 m/km
(3) 120 m/km
(2) 40 m/km
(4) 160 m/km

9. According to the *Earth Science Reference Tables*, if atmospheric pressure measurements were taken at regular intervals from sea level to the stratopause, the measurements would most likely show that the pressure
   (1) decreases, only
   (2) increases, only
   (3) remains the same
   (4) decreases, then increases

10. The diagram below shows a view of the Earth in space with respect to the Sun's rays. Point *X* is a location on the Earth's surface.

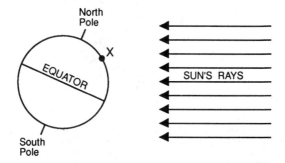

   Which event would occur at point *X* on the date represented in the diagram?
   (1) the beginning of the winter season
   (2) the formation of the longest noontime shadows for the year
   (3) the lowest noontime altitude of the Sun for the year
   (4) the greatest duration of insolation for the year

11. The diagram below shows the noontime shadow cast by a vertical post located in New York State.

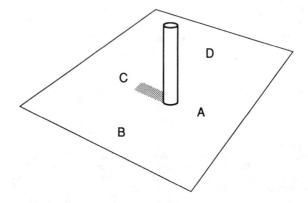

   Which letter indicates a location south of the post?
   (1) *A*   (3) *C*
   (2) *B*   (4) *D*

12. Which statement best explains the apparent daily motion of the stars around Polaris?
   (1) The Earth's orbit is an ellipse.
   (2) The Earth has the shape of an oblate spheroid.
   (3) The Earth rotates on its axis.
   (4) The Earth revolves around the Sun.

13. According to the *Earth Science Reference Tables,* which material requires the *least* amount of energy to change its temperature 1 C° per gram?
    (1) iron
    (2) ice
    (3) water
    (4) lead

14. During which month does the Sun rise north of due east in New York State?
    (1) February
    (2) July
    (3) October
    (4) January

15. A piece of a plant in a classroom fishtank moved upward and across the tank, away from the water heater. When the plant reached the other side of the tank, it sank before moving back toward the heater. What type of energy transfer does this movement represent?
    (1) convection
    (2) conduction
    (3) refraction
    (4) radiation

16. Which type of surface absorbs the greatest amount of electromagnetic energy from the Sun?
    (1) smooth, shiny, and dark in color
    (2) rough, dull, and dark in color
    (3) smooth, shiny, and light in color
    (4) rough, dull, and light in color

17. The curving path of planetary winds is caused by
    (1) the Earth's revolution
    (2) the Earth's rotation
    (3) ocean currents
    (4) weather fronts

18. Water vapor and carbon dioxide affect the warming of the Earth's atmosphere because they both
    (1) have high specific heats
    (2) scatter insolation
    (3) absorb infrared radiation
    (4) reflect ultraviolet radiation

19. On a July afternoon in New York State, the barometric pressure is 29.85 inches and falling. This reading most likely indicates
    (1) an approaching storm
    (2) rapidly clearing skies
    (3) continuing fair weather
    (4) gradually improving conditions

20. The data table below shows average daily air temperature, windspeed, and relative humidity for 4 days at a single location.

Day	Air Temperature (°F)	Windspeed (mph)	Relative Humidity (%)
Monday	40	15	60
Tuesday	65	10	75
Wednesday	80	20	30
Thursday	85	0	95

On which day was the air closest to being saturated with water vapor?
    (1) Monday
    (2) Tuesday
    (3) Wednesday
    (4) Thursday

21. What is the approximate dewpoint temperature when the dry-bulb temperature is 2°C and the wet-bulb temperature is 0°C?
    (1) −1°C
    (2) −2°C
    (3) −3°C
    (4) −6°C

22. Equal quantities of water are placed in four uncovered containers with different shapes and left on a table at room temperature. From which container will the water evaporate most rapidly?

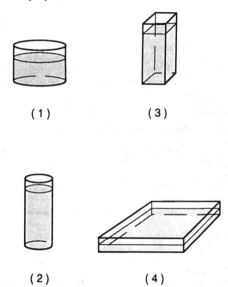

(1)

(3)

(2)

(4)

23. A weather map of New York State shows isobars that are close together, indicating a steep pressure gradient. Which weather condition is most likely present?
    (1) dry air
    (2) strong winds
    (3) low temperatures
    (4) low visibility

24. Which event usually occurs when air is cooled to its dewpoint temperature?
    (1) freezing
    (2) evaporation
    (3) condensation
    (4) transpiration

25. Through which sediment does water infiltrate slowly? [Refer to the *Earth Science Reference Tables.*]
    (1) sand
    (2) silt
    (3) clay
    (4) pebbles

26. In the middle latitudes of the Southern Hemisphere, the warmest month is usually
    (1) February
    (2) April
    (3) July
    (4) October

27. According to the *Earth Science Reference Tables,* at which of these latitudes would average annual precipitation be greatest?
    (1) 0°
    (2) 30° N
    (3) 90° N
    (4) 90° S

28. In which type of climate does the greatest amount of chemical weathering of rock occur?
    (1) cold and dry
    (2) cold and moist
    (3) warm and dry
    (4) warm and moist

29. Four different kinds of particles (A, B, C, and D) with the same shape and diameter were mixed and poured into a column of water. The mass, volume, and density of the particles are shown below.

Particle	Mass (g)	Volume (cm^3)	Density (g/cm^3)
A	100	67	1.5
B	100	33	3.0
C	100	22	4.5
D	100	17	6.0

Which diagram best shows how the particle beds would be arranged in the column of water after settling?

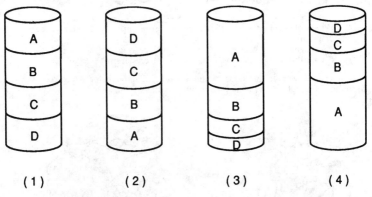

(1)   (2)   (3)   (4)

30. Which conditions are most likely to develop over a land area next to an ocean during a hot, sunny afternoon?
    (1) The air temperature over the land is lower than the air temperature over the ocean, and a breeze blows from the land.
    (2) The air temperature over the land is higher than the air temperature over the ocean, and a breeze blows from the land.
    (3) The air pressure over the land is higher than the air pressure over the ocean, and a breeze blows from the ocean.
    (4) The air pressure over the land is lower than the air pressure over the ocean, and a breeze blows from the ocean.

31. On the Earth's surface, transported materials are more common than residual materials. This condition is mainly the result of
    (1) subduction   (3) folding
    (2) erosion   (4) recrystallization

32. How are dissolved materials carried in a river?
    (1) in solution   (3) by precipitation
    (2) in suspension   (4) by bouncing and rolling

33. A state of dynamic equilibrium exists in a stream when the
    (1) rate of stream flow is gradually decreasing
    (2) rate of stream flow is gradually increasing
    (3) rate of erosion is greater than the rate of deposition
    (4) rates of erosion and deposition are the same

34. The diagrams below represent samples of five different minerals found in the rocks of the Earth's crust.

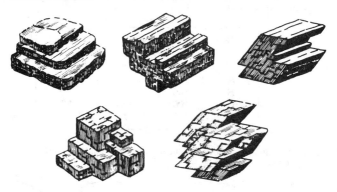

Which physical property of minerals is represented by the flat surfaces in the diagrams?
(1) magnetism   (3) cleavage
(2) hardness   (4) crystal size

Base your answers to questions 35 and 36 on the diagram below which represents the formation of a sedimentary rock. [Sediments are drawn actual size.]

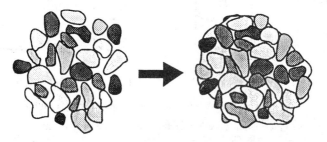

Sediments   Sedimentary Rock

35. The formation of which sedimentary rock is shown in the diagram?
(1) conglomerate   (3) siltstone
(2) sandstone   (4) shale

36. Which two processes formed this rock?
(1) folding and faulting   (3) compaction and cementation
(2) melting and solidification   (4) heating and application of pressure

37. The geologic cross section below represents an igneous intrusion into layers of rock strata.

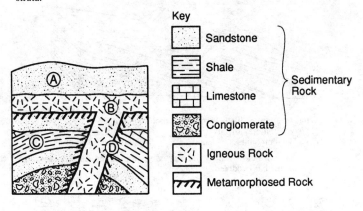

Key
- Sandstone
- Shale
- Limestone
- Conglomerate
} Sedimentary Rock
- Igneous Rock
- Metamorphosed Rock

Which letter indicates a point where unmelted rock changed as a result of increased temperature and pressure?
(1) *A*   (3) *C*
(2) *B*   (4) *D*

38. Rock *X* and rock *Y* are igneous rocks with identical mineral composition. Rock *X* has no visible crystals and rock *Y* has large, visible crystals. What can be inferred about rock *Y*?
(1) It cooled at the Earth's surface, more slowly than rock *X*.
(2) It cooled beneath the Earth's surface, more slowly than rock *X*.
(3) It cooled at the Earth's surface, more quickly than rock *X*.
(4) It cooled beneath the Earth's surface, more quickly than rock *X*.

39. How does the oceanic crust compare to the continental crust?
(1) The oceanic crust is thinner and contains less basalt.
(2) The oceanic crust is thinner and contains more basalt.
(3) The oceanic crust is thicker and contains less basalt.
(4) The oceanic crust is thicker and contains more basalt.

40. The structure of the Earth's interior is best inferred by
(1) analyzing worldwide seismic data
(2) measuring crustal temperature ranges
(3) determining crustal density differences
(4) observing rock samples from surface bedrock

41. Which inference is supported by a study of the Earth's magnetic rock record?
(1) The Earth's magnetic field is only 2 million years old.
(2) The Earth's magnetic field is 50 times stronger now than in the past.
(3) The Earth's magnetic poles are usually located at 0° latitude.
(4) The Earth's magnetic poles appear to have changed location over time.

42. According to the *Earth Science Reference Tables,* during which period were North America, Africa, and South America closest?
(1) Tertiary   (3) Triassic
(2) Cretaceous   (4) Ordovician

43. The block diagram below represents a geologic cross section of a mountain range.

What action most likely formed this mountain range?
(1) contact metamorphism   (3) volcanic eruptions
(2) glacial erosion   (4) earthquake faulting

44. The diagram below represents a cross section of the Earth's crust in which no overturning has occurred.

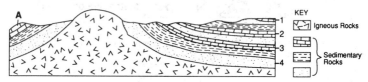

Which numbered rock layer on the right is most likely the same age as layer A on the left?
(1) 1   (3) 3
(2) 2   (4) 4

45. Geologic time is divided into specific periods and epochs based on
(1) fossil evidence of Earth organisms
(2) inferred positions of Earth landmasses
(3) rock types found in mountainous areas
(4) uplift and erosion of New York State bedrock

46. According to the *Earth Science Reference Tables,* during which geologic time period did the salt and gypsum deposits near Syracuse form?
(1) Cambrian   (3) Silurian
(2) Ordovician   (4) Devonian

47. The age of a Moon rock can be found by analyzing a sample to compare the relative amounts of
(1) U^{238} and Pb^{206} (3) C^{14} and Pb^{206}
(2) U^{238} and C^{14} (4) U^{238} and Sr^{87}

48. Theories of evolution suggest that variations between members of the same species give the species greater probability of
(1) remaining unchanged   (3) becoming fossilized
(2) surviving environmental changes   (4) becoming extinct

49. The cartoon below represents the time of the last dinosaurs and the earliest mammals.

According to the *Earth Science Reference Tables*, the cartoon could represent the boundary between which two units of geologic history?
(1) Archean and Proterozoic
(2) Precambrian and Paleozoic
(3) Ordovician and Silurian
(4) Mesozoic and Cenozoic

50. The geologic cross section below shows an unconformity between gneiss and the Cambrian-age Potsdam sandstone in northern New York State.

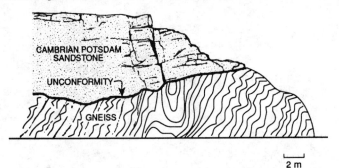

According to the *Earth Science Reference Tables*, what is the most probable age of the gneiss at this location?
(1) Precambrian
(2) Silurian
(3) Ordovician
(4) Cretaceous

51. The diagram below represents the surface topography of a mountain valley.

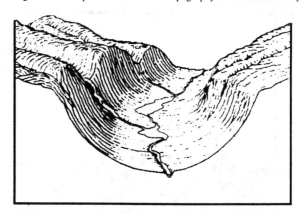

Which agent of erosion most likely created the shape of the valley shown in the diagram?
(1) wind
(2) glaciers
(3) ocean waves
(4) running water

52. If weathering and erosion were the only geological processes taking place on the Earth, most landscapes would be characterized by
(1) low relief and gentle gradients
(2) low relief and steep gradients
(3) high relief and gentle gradients
(4) high relief and steep gradients

53. According to the *Earth Science Reference Tables,* which New York State landscape region has intensely metamorphosed surface bedrock?
(1) Appalachian Plateau
(2) Adirondack Mountains
(3) Atlantic Coastal Plain
(4) Erie-Ontario Lowlands

Base your answers to questions 54 and 55 on the map below which shows present drainage basins in New York State.

54. In the Susquehanna-Chesapeake drainage basin, rivers flow over bedrock composed primarily of
    (1) gneisses, quartzites, and marbles
    (2) shales, slates, and diabases
    (3) conglomerates, granites, and rhyolites
    (4) limestones, shales, and sandstones

55. In which drainage basin are the Finger Lakes located? [Refer to the *Earth Science Reference Tables.*]
    (1) Mohawk–Hudson
    (2) Susquehanna–Chesapeake
    (3) Ontario–St. Lawrence
    (4) Champlain–St. Lawrence

PART II

This part consists of ten groups, each containing five questions. Choose seven of these ten groups. Be sure that you answer all five questions in each group chosen. [35]

Group 1

If you choose this group, be sure to answer questions 56-60.

Base your answers to questions 56 through 60 on the latitude and longitude system shown below and your knowledge of Earth science. The map represents a part of the Earth's surface and its latitude-longitude coordinates. Points *A* through *F* represent locations in this area.

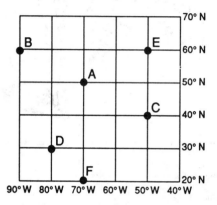

56. What is the compass direction from point *D* toward point *E*?
    (1) southwest   (3) northwest
    (2) southeast   (4) northeast

57. Points *B* and *E* would *not* have the same value for measurements of
    (1) latitude   (3) duration of solar day
    (2) time   (4) altitude of the Sun at solar noon

58. Which location would have the greatest angle of insolation on June 21?
    (1) *A*   (3) *C*
    (2) *E*   (4) *F*

59. How are latitude and longitude lines drawn on a globe of the Earth?
    (1) Latitude lines are parallel and longitude lines meet at the poles.
    (2) Latitude lines are parallel and longitude lines meet at the Equator.
    (3) Longitude lines are parallel and latitude lines meet at the poles.
    (4) Longitude lines are parallel and latitude lines meet at the Equator.

Note that question 60 has only three choices.

60. As a person travels from location *B* to location *E*, the observed altitude of Polaris will
    (1) decrease
    (2) increase
    (3) remain the same

Group 2

If you choose this group, be sure to answer questions 61–65.

Base your answers to questions 61 through 65 on the *Earth Science Reference Tables,* the diagram below, and your knowledge of Earth science. The diagram represents the orbits of three planets (X, Y, and Z) around star A. Star A is located at one focus and point B is the other focus. Numbers 1 through 9 represent different positions of the three planets. The arrows show the direction of revolution.

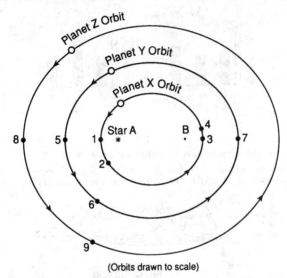

(Orbits drawn to scale)

61. The orbital paths of these planets around star A can best be described as having
    (1) the same period of rotation
    (2) major axes of the same length
    (3) an elliptical shape, with star A at one focus
    (4) a circular shape, with star A at one focus

62. The time required for planet X to travel from point 1 to point 2 is approximately the same as the time required for planet X to travel from point
    (1) 2 to point 3   (3) 2 to point 4
    (2) 3 to point 4   (4) 3 to point 1

63. Which statement about the period of revolution of the planets is correct?
    (1) The planets have equal periods of revolution.
    (2) Planet X has a longer period of revolution than planet Y.
    (3) Planet Y has a longer period of revolution than planet Z.
    (4) Planet Z has a longer period of revolution than planet X.

64. Which number indicates the position at which a planet would have the greatest gravitational attraction to star A? [Assume that all three planets have the same mass.]
    (1) 7   (3) 3
    (2) 6   (4) 5

65. Which graph best represents the eccentricity of the orbits of planets *X*, *Y*, and *Z*?

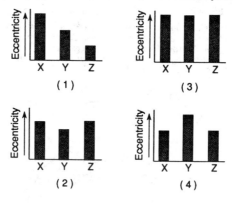

Group 3

If you choose this group, be sure to answer questions 66–70.

Base your answers to questions 66 through 70 on the *Earth Science Reference Tables,* the diagram below which represents the water cycle, and your knowledge of Earth science.

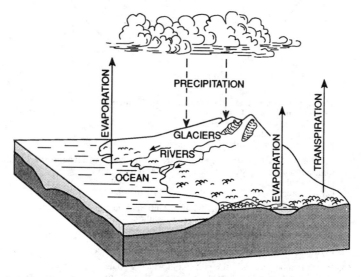

66. By which process does most water vapor enter the atmosphere?
    (1) evaporation from lakes and rivers
    (2) evaporation from ocean surfaces
    (3) evapotranspiration from land areas
    (4) sublimation from polar ice and snow

67. The small arrows drawn near the rivers represent the direction of
    (1) capillarity
    (2) runoff
    (3) absorption
    (4) infiltration

68. During which process does water vapor release 540 calories of latent heat per gram?
    (1) condensation
    (2) evaporation
    (3) transpiration
    (4) precipitation

69. Precipitation is most likely occurring at the time represented in the diagram because
    (1) the air has been warmed due to expansion
    (2) no condensation nuclei are present in the air
    (3) the relative humidity of the air is low
    (4) the water droplets are heavy enough to fall

70. Water will enter the soil if the ground surface is
    (1) impermeable and saturated
    (2) impermeable and unsaturated
    (3) permeable and saturated
    (4) permeable and unsaturated

Group 4

If you choose this group, be sure to answer questions 71–75.

Base your answers to questions 71 through 75 on the *Earth Science Reference Tables*, the weather map below, and your knowledge of Earth science. The map shows part of the southern United States and northern Mexico.

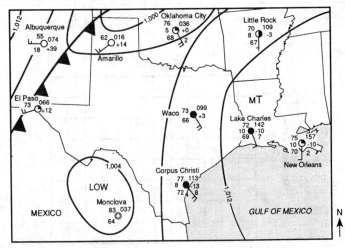

71. At which city is the visibility 8 miles?
    (1) Little Rock, Arkansas
    (2) Lake Charles, Louisiana
    (3) Oklahoma City, Oklahoma
    (4) New Orleans, Louisiana

72. The isolines on this map connect locations that have the same
    (1) dewpoint temperature
    (2) air temperature
    (3) barometric pressure
    (4) relative humidity

73. Which kind of air mass is influencing the weather of Lake Charles, Louisiana?
    (1) warm and dry
    (2) warm and moist
    (3) cold and dry
    (4) cold and moist

74. Southeast winds at 20 knots are occurring at
    (1) Albuquerque, New Mexico
    (2) Amarillo, Texas
    (3) Oklahoma City, Oklahoma
    (4) El Paso, Texas

75. Which city has the *least* chance of precipitation during the next 3 hours?
    (1) Oklahoma City, Oklahoma
    (2) Waco, Texas
    (3) Lake Charles, Louisiana
    (4) Albuquerque, New Mexico

Group 5

If you choose this group, be sure to answer questions 76–80.

Base your answers to questions 76 through 80 on the table below and your knowledge of Earth science. The table gives water budget data for Syracuse, New York. A blank water budget graph is included for your use. The values shown are in millimeters of water.

Water Budget for Syracuse, NY

Month		J	F	M	A	M	J	J	A	S	O	N	D	Yearly Totals
Precipitation	P	72	68	81	75	74	87	84	82	72	76	68	72	911
Potential Evapotranspiration	E_p	0	0	3	34	83	115	134	122	84	46	15	0	636
	$P - E_p$	72	68	78	41	−9	−28	−50	−40	−12	30	53	72	— —
Change in Storage	ΔSt	0	0	0	0	−9	−28	−50	−13	0	30	53	17	— —
Storage	St	100	100	100	100	91	63	13	0	0	30	83	100	— —
Actual Evapotranspiration	E_a	0	0	3	34	83	115	134	95	72	46	15	0	597
Deficit	D	0	0	0	0	0	0	0	27	12	0	0	0	39
Surplus	S	72	68	78	41	0	0	0	0	0	0	0	55	314

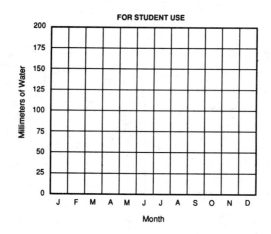

FOR STUDENT USE

76. Which graph best represents the yearly precipitation (P) and potential evapotranspiration (E_p) for Syracuse, New York?

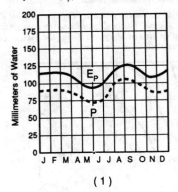

(1)

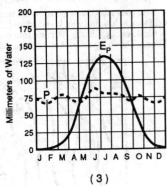

(3)

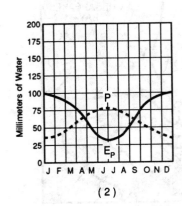

(2)

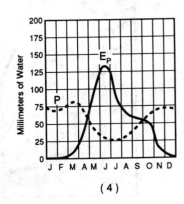

(4)

77. What is the approximate average monthly precipitation for Syracuse?
    (1) 70 mm
    (2) 76 mm
    (3) 80 mm
    (4) 85 mm

78. A moisture surplus exists when the
    (1) storage is 0 and precipitation equals potential evapotranspiration
    (2) storage is 100 and precipitation is greater than potential evapotranspiration
    (3) deficit is 0 and the actual evapotranspiration equals the potential evapotranspiration
    (4) deficit is 100 and the actual evapotranspiration is greater than the potential evapotranspiration

79. What is the total amount of moisture in storage in the month of May?
    (1) 9 mm
    (2) 34 mm
    (3) 63 mm
    (4) 91 mm

80. In which month is ground-water usage greatest?
    (1) March
    (2) July
    (3) August
    (4) November

Group 6

If you choose this group, be sure to answer questions 81–85.

Base your answers to questions 81 through 85 on the *Earth Science Reference Tables,* the field map below, and your knowledge of Earth science. The field map shows the average size of particles deposited by streams that drained an area of Maryland during the Pleistocene Epoch. The field values represent particle diameters in centimeters.

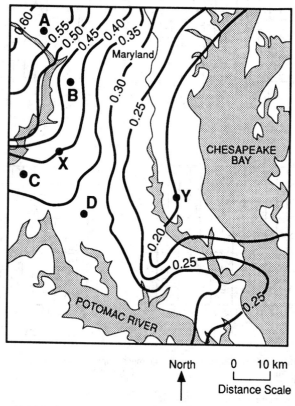

81. At which location would the sediment particles have an average diameter of 0.33 centimeter?
    (1) *A*   (3) *C*
    (2) *B*   (4) *D*

82. Which particle size would be most common at location *X*? [Particles are drawn actual size.]

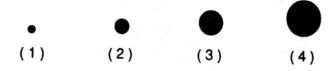

83. The sedimentary deposits shown on this field map would be classified as
    (1) silt
    (2) sand
    (3) pebbles
    (4) cobbles

84. Particles of sediment collected at location Y contain intergrown crystals of quartz, potassium feldspar, and hornblende. From which rock did these sediments most likely weather?
    (1) granite
    (2) gabbro
    (3) sandstone
    (4) limestone

85. The diagrams below represent magnified views of rock particles. Which rock particle was most likely transported by these streams for the longest period of time?

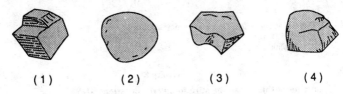

(1)   (2)   (3)   (4)

Group 7

If you choose this group, be sure to answer questions 86–90.

Base your answers to questions 86 through 90 on the *Earth Science Reference Tables,* the table below, and your knowledge of Earth science. The table shows some of the data collected at two seismic stations, A and B. Some data have been omitted.

Station	Arrival Time of P-Wave	Arrival Time of S-Wave	Difference in Arrival Times of P- and S-Waves	Distance to Epicenter
A	6:02:00 p.m.	6:07:30 p.m.	5 min 30 sec	— km
B	— p.m.	6:11:20 p.m.	7 min 20 sec	5,700 km

86. Which seismogram most accurately represents the arrival of the *P*- and *S*-waves at station *A*?

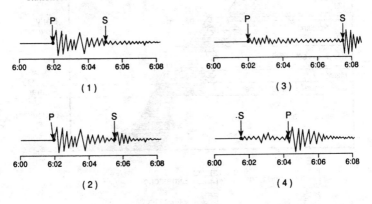

87. What is the approximate distance from the epicenter to station *A*?
    (1) 1,400 km   (3) 3,000 km
    (2) 1,900 km   (4) 4,000 km

88. Which statement best describes the seismic waves received at station *B*?
    (1) The *P*-wave arrived at 6:12 p.m.
    (2) The *S*-wave arrived before the *P*-wave.
    (3) The *P*-wave had the greatest velocity.
    (4) The *S*-wave passed through a fluid before reaching station *B*.

89. What is the minimum number of *additional* stations from which scientists must collect data in order to locate the epicenter of this earthquake?
    (1) 1   (3) 3
    (2) 2   (4) 0

90. What was the origin time of this earthquake?
    (1) 5:55:00 p.m.   (3) 6:06:00 p.m.
    (2) 6:00:00 p.m.   (4) 6:11:20 p.m.

Group 8

If you choose this group, be sure to answer questions 91–95.

Base your answers to questions 91 through 95 on the *Earth Science Reference Tables*, the diagram below and table on the next page, and your knowledge of Earth science. The diagram represents a cross section of a hillside in eastern New York State. The names of the rock layers and the geologic time periods in which they were formed are given. Layer thicknesses are shown in feet. The table shows the types of fossils found in the rock layers shown. A checkmark (✓) indicates that the fossil is present.

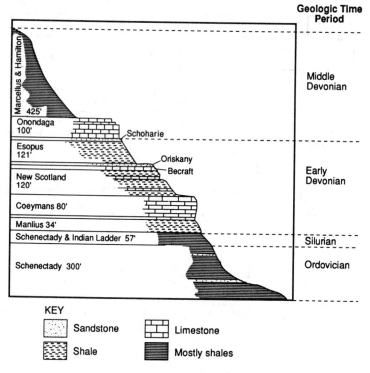

ROCK LAYER	FOSSILS													
	Graptolites	Crinoids	Starfish	Brachiopods	Worms and/or Worm Trails	Pelecypods	Gastropods	Cephalopods	Trilobites	Ostracods	Eurypterids	Bryozoans	Corals	Sponges
Hamilton				✓	✓	✓	✓							
Marcellus				✓				✓						
Onondaga				✓		✓	✓	✓	✓			✓	✓	
Schoharie				✓		✓	✓	✓	✓			✓	✓	
Esopus					✓									
Oriskany				✓			✓							
Becraft		✓											✓	
New Scotland		✓		✓		✓	✓	✓	✓			✓	✓	✓
Coeymans		✓		✓										
Manlius		✓		✓		✓				✓		✓	✓	
Indian Ladder	✓	✓		✓					✓			✓		
Schenectady	✓	✓	✓	✓	✓	✓	✓	✓	✓	✓				

91. Which rock layers are the youngest?
   (1) Schenectady and Indian Ladder
   (2) Esopus and Schoharie
   (3) Manlius and Coeymans
   (4) Marcellus and Hamilton

92. Which fossil type could be used as an index fossil for the Schenectady layers?
   (1) sponges
   (2) eurypterids
   (3) brachiopods
   (4) crinoids

93. Which two rock layers seem to be most resistant to erosion?
   (1) Hamilton and Manlius
   (2) Esopus and New Scotland
   (3) Coeymans and Onondaga
   (4) Becraft and Schenectady

94. What do the Marcellus and Hamilton layers have in common with the Schenectady and Indian Ladder layers?
   (1) They developed during the same time period.
   (2) They are composed mostly of shales.
   (3) They formed directly over limestone.
   (4) They have the same thickness.

95. Crinoids and sponges are collected from one of these rock layers. During which time period did this rock layer form?
   (1) Silurian
   (2) Late Ordovician
   (3) Early Devonian
   (4) Triassic

Group 9

If you choose this group, be sure to answer questions 96–100.

Base your answers to questions 96 through 100 on the *Earth Science Reference Tables,* the map below, and your knowledge of Earth science. The map shows geologic features of Cape Cod, Massachusetts. The locations of several towns are shown as **O** .

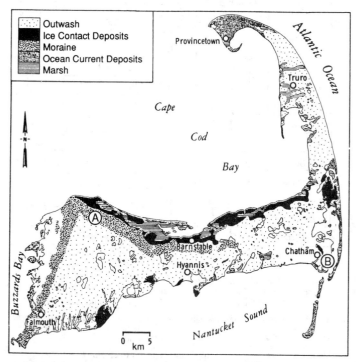

96. What is the approximate straight-line distance in kilometers from Hyannis to Chatham?
    (1) 5 km
    (2) 15 km
    (3) 21 km
    (4) 25 km

97. The unsorted sediments at location *A* were deposited during the advance and retreat of the last continental ice sheet. During which geologic epoch were these sediments deposited?
    (1) Pleistocene
    (2) Pliocene
    (3) Miocene
    (4) Oligocene

98. Soil and water samples taken from various locations show significant amounts of pollutants. Which statement best supports these findings?
    (1) The effects of pollution have determined the stream patterns of this landscape.
    (2) Soil deposited by glaciers is generally polluted.
    (3) The activities of humans have altered the environment.
    (4) Most glacial deposits are impermeable.

99. Ice contact deposits are found where sediment was deposited directly against glacial ice. The locations of the ice contact deposits represented on this map were probably determined by
    (1) direct visual observations of the glacier that formed Cape Cod
    (2) inferences made from observations of the landscapes
    (3) classifying meltwater samples from various locations
    (4) studying pollutants in the ground water at Provincetown

100. Which landscape region of New York State contains depositional and landscape features most similar to those of Cape Cod?
    (1) Allegheny Plateau   (3) Atlantic Coastal Plain
    (2) Adirondack Mountains   (4) Tug Hill Plateau

Group 10

If you choose this group, be sure to answer questions 101–105.

Base your answers to questions 101 through 105 on the *Earth Science Reference Tables* and your knowledge of Earth science.

101. An air pressure of 1,005 millibars is equivalent to approximately how many inches of mercury?
    (1) 29.58   (3) 29.68
    (2) 29.62   (4) 29.72

102. A student determines the density of a rock to be 2.2 grams per cubic centimeter. If the accepted density of the rock is 2.5 grams per cubic centimeter, what is the percent deviation (percentage of error) from the accepted value?
    (1) 8.8%   (3) 13.6%
    (2) 12.0%   (4) 30.0%

103. Water has its greatest density at a temperature of
    (1) −6°C   (3) 32°C
    (2) 10°C   (4) 4°C

104. Which process could lead most directly to the formation of a sedimentary rock?
    (1) metamorphism of unmelted material
    (2) slow solidification of molten material
    (3) sudden upwelling of lava at a mid-ocean ridge
    (4) precipitation of minerals from evaporating water

105. In which New York State landscape region is the location 42° North latitude by 74°30′ West longitude found?
    (1) the Catskills   (3) Hudson-Mohawk Lowlands
    (2) Adirondack Mountains   (4) Taconic Mountains

EARTH SCIENCE

June 18, 1996

PART I

Answer all 55 questions in this part.   [55]

Directions (1–55): For *each* statement or question, select the word or expression that, of those given, best completes the statement or answers the question.

1. Which graph most likely illustrates a cyclic change?

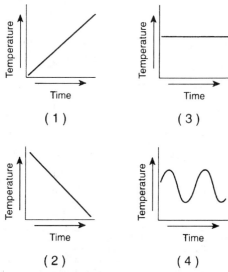

2. Which statement about an unidentified rock sample is most likely an inference?
   (1) The rock is composed of large crystals.
   (2) The rock has shiny, wavy mineral bands.
   (3) The rock is a metamorphic rock.
   (4) The rock has no visible fossils.

3. A student determined the porosity of a sample of soil to be 37.6%. The actual porosity is 42.3%. The student's percent deviation from the accepted value (percentage of error) is approximately
   (1) 4.7%   (3) 12.5%
   (2) 11.1%   (4) 79.9%

4. The diagrams below represent two solid objects, *A* and *B,* with different densities.

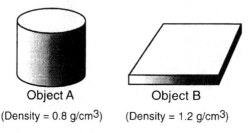

Object A
(Density = 0.8 g/cm³)

Object B
(Density = 1.2 g/cm³)

What will happen when the objects are placed in a container of water (water temperature = 4°C)?
(1) Both objects will sink.
(2) Both objects will float.
(3) Object *A* will float, and object *B* will sink.
(4) Object *B* will float, and object *A* will sink.

5. Measurements of the Sun's altitude at the same time from two different Earth locations a known distance apart are often used to determine the
(1) circumference of the Earth
(2) period of the Earth's revolution
(3) length of the major axis of the Earth's orbit
(4) eccentricity of the Earth's orbit

6. In which atmospheric layer is most water vapor found?
(1) troposphere
(2) stratosphere
(3) mesosphere
(4) thermosphere

7. Which diagram most accurately shows the cross-sectional shape of the Earth?

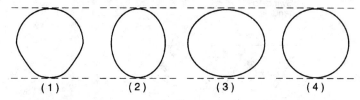

    (1)   (2)   (3)   (4)

8. The diagram below shows an instrument made from a drinking straw, protractor, string, and rock.

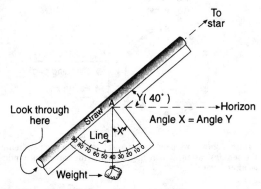

This instrument was most likely used to measure the
(1) distance to a star
(2) altitude of a star
(3) mass of the Earth
(4) mass of the suspended weight

9. In New York State, solar noon occurs each day when the
(1) Sun is at its highest altitude
(2) Sun is directly overhead
(3) clock reads 12 o'clock noon
(4) observer's shadow is longest

10. In New York State, which day has the shortest duration of insolation?
(1) March 21
(2) June 21
(3) September 21
(4) December 21

11. All objects warmer than 0 Kelvin (absolute zero) must be
    (1) radiating electromagnetic energy
    (2) condensing to form a gas
    (3) warmer than 0° Celsius
    (4) expanding in size

Base your answers to questions 12 and 13 on the diagrams below, which represent two views of a swinging Foucault pendulum with a ring of 12 pegs at its base.

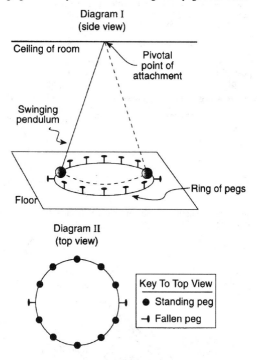

12. Diagram II shows two pegs tipped over by the swinging pendulum at the beginning of the demonstration. Which diagram shows the pattern of standing pegs and fallen pegs after several hours?

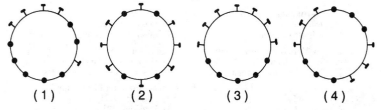

13. The predictable change in the direction of swing of a Foucault pendulum provides evidence that the
    (1) Sun rotates on its axis
    (2) Sun revolves around the Earth
    (3) Earth rotates on its axis
    (4) Earth revolves around the Sun

14. Which angle of the Sun above the horizon produces the greatest intensity of sunlight per unit area?
    (1) 25°   (3) 60°
    (2) 40°   (4) 70°

15. A city located near the center of a large continent has colder winters and warmer summers than a city at the same elevation and latitude located on the continent's coast. Which statement best explains the difference between the cities' climates?
    (1) Windspeeds are greater over land than over oceans.
    (2) Air masses originate only over land.
    (3) Land has a lower specific heat than water.
    (4) Water changes temperature more rapidly than land.

16. The greatest amount of energy would be gained by 1,000 grams of water when it changes from
    (1) water vapor to liquid water   (3) liquid water to ice
    (2) liquid water to water vapor   (4) ice to liquid water

17. Which diagram shows the position of the Earth relative to the Sun's rays during a winter day in the Northern Hemisphere?

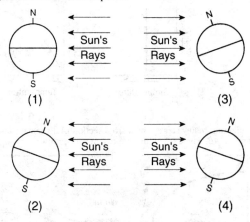

18. According to the *Earth Science Reference Tables,* an air pressure of 30.21 inches of mercury is equal to approximately
    (1) 1,015 mb   (3) 1,020 mb
    (2) 1,017 mb   (4) 1,023 mb

19. The diagram below represents the path of visible light as it travels from air to water to air through a glass container of water.

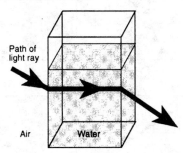

The light did *not* travel in a straight line because of
(1) convection
(2) scattering
(3) absorption
(4) refraction

20. Which graph best represents the relationship between air temperature and air density in the atmosphere?

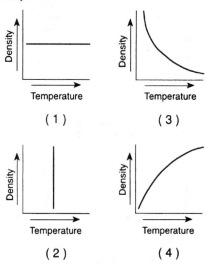

21. The diagram below represents a cross section of air masses and frontal surfaces along line *AB*. The dashed lines represent precipitation.

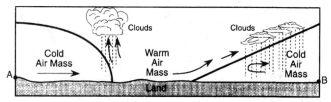

Which weather map best represents this frontal system?

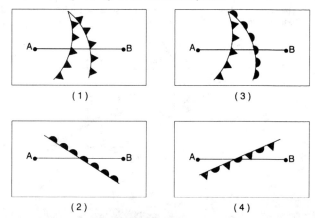

22. In the Northern Hemisphere, in which direction does surface wind circulate in a high-pressure air mass?
    (1) clockwise and toward the center
    (2) clockwise and away from the center
    (3) counterclockwise and toward the center
    (4) counterclockwise and away from the center

23. The greatest source of moisture entering the atmosphere is evaporation from the surface of
    (1) the oceans
    (2) the land
    (3) lakes and streams
    (4) ice sheets and glaciers

24. Most of the Earth's surface ocean currents are caused by
    (1) stream flow from continents
    (2) differences in ocean water density
    (3) the revolution of the Earth
    (4) the prevailing winds

25. In order for clouds to form, cooling air must be
    (1) saturated and have no condensation nuclei
    (2) saturated and have condensation nuclei
    (3) unsaturated and have no condensation nuclei
    (4) unsaturated and have condensation nuclei

26. The cross section below shows several locations in the state of Washington and annual precipitation at each location. The arrows represent the prevailing wind direction.

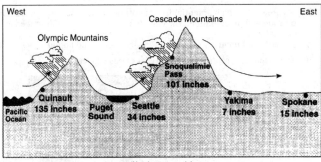

(Not drawn to scale)

Why do the windward sides of these mountain ranges receive more precipitation than the leeward sides?
    (1) Rising air expands and cools.
    (2) Rising air compresses and cools.
    (3) Sinking air expands and cools.
    (4) Sinking air compresses and cools.

27. The diagram below represents zones within soil and rock. The zones are determined by the kinds of movement or lack of movement of water occurring within them.
    What is the deepest zone into which water can be pulled by gravity?
    (1) aerated zone
    (2) capillary fringe
    (3) saturated zone
    (4) impermeable zone

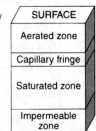

28. According to the water budget graph below, during which month will the soil moisture most likely be depleted at this location?

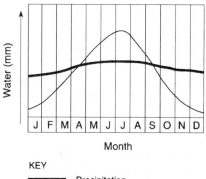

KEY

—— Precipitation

–––– Potential evapotranspiration

(1) February   (3) August
(2) April   (4) October

29. An area with a high potential for evapotranspiration has little actual evapotranspiration and precipitation. The climate of this area is best described as
    (1) hot and arid   (3) cold and arid
    (2) hot and humid   (4) cold and humid

30. The map below represents a river as it enters a lake.

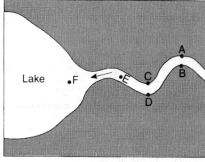

At which locations is the amount of deposition greater than the amount of erosion?
(1) A,C, and E   (3) B, D, and F
(2) B, C, and F   (4) A, D, and E

31. What change will a pebble usually undergo when it is transported a great distance by streams?
    (1) It will become jagged and its mass will decrease
    (2) It will become jagged and its volume will increase.
    (3) It will become rounded and its mass will increase.
    (4) It will become rounded and its volume will decrease.

32. The velocity of a stream is decreasing. As the velocity approaches zero, which size particle will most likely remain in suspension?
    (1) clay   (3) pebble
    (2) sand   (4) boulder

33. A large, scratched boulder is found in a mixture of unsorted, smaller sediments forming a hill in central New York State. Which agent of erosion most likely transported and then deposited this boulder?
(1) wind
(2) glacier
(3) ocean waves
(4) running water

34. Which element in the Earth's crust makes up the largest volume of most minerals?
(1) oxygen
(2) nitrogen
(3) hydrogen
(4) iron

35. A fine-grained igneous rock was probably formed by
(1) weathering and erosion
(2) great heat and pressure that did not produce melting
(3) rapid cooling of molten material
(4) burial and cementation of sediment

36. Which two igneous rocks could have the same mineral composition?
(1) rhyolite and diorite
(2) pumice and scoria
(3) peridotite and andesite
(4) gabbro and basalt

37. Which diagram best represents a sample of the metamorphic rock gneiss? (Diagrams show actual size.)

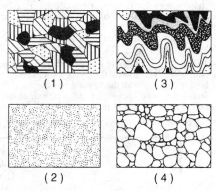

(1)   (3)

(2)   (4)

38. A seismogram recorded at a seismic station is shown below.

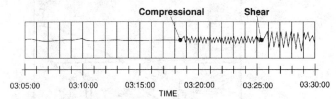

Which information can be determined by using this seismogram?
(1) the depth of the earthquake's focus
(2) the direction of the earthquake's focus
(3) the location of the earthquake's epicenter
(4) the distance to the earthquake's epicenter

39. Which statement best describes the relationship between the travel rates and travel times of earthquake P-waves and S-waves from the focus of an earthquake to a seismograph station? [Refer to the *Earth Science Reference Tables.*]
(1) P-waves travel at slower rate and take less time.
(2) P-waves travel at a faster rate and take less time.
(3) S-waves travel at a slower rate and take less time.
(4) S-waves travel at a faster rate and take less time.

40. Where is the thickest part of the Earth's crust?
    (1) at the edge of continental shelves
    (2) at mid-ocean ridges
    (3) under continental mountain ranges
    (4) under volcanic islands

41. Differences in hardness between minerals are most likely caused by the
    (1) internal arrangement of atoms
    (2) external arrangement of flat surfaces
    (3) number of pointed edges
    (4) number of cleavage planes

42. According to the *Earth Science Reference Tables,* the rock between 2,900 kilometers and 5,200 kilometers below the Earth's surface is inferred to be
    (1) an iron-rich solid   (3) a silicate-rich solid
    (2) an iron-rich liquid   (4) a silicate-rich liquid

43. An earthquake's *P*-wave traveled 4,800 kilometers and arrived at a seismic station at 5:10 p.m. At approximately what time did the earthquake occur?
    (1) 5:02 p.m.   (3) 5:10 p.m.
    (2) 5:08 p.m.   (4) 5:18 p.m.

44. The theory of continental drift suggests that the
    (1) continents moved due to changes in the Earth's orbital velocity
    (2) continents moved due to the Coriolis effect, caused by the Earth's rotation
    (3) present-day continents of South America and Africa are moving toward each other
    (4) present-day continents of South America and Africa once fit together like puzzle parts

45. What is the relative age of a fault that cuts across many rock layers?
    (1) The fault is younger than all the layers it cuts across.
    (2) The fault is older than all the layers it cuts across.
    (3) The fault is the same age as the top layer it cuts across.
    (4) The fault is the same age as the bottom layer it cuts across.

46. The cartoon below is a humorous look at geologic history.

Early Pleistocene mermaids

If Early Pleistocene mermaids had existed, their fossil remains would be the same age as fossils of
(1) armored fish
(3) trilobites
(2) mastodons
(4) dinosaurs

47. In order for an organism to be used as an index fossil, the organism must have been geographically widespread and must have
(1) lived on land
(3) been preserved by volcanic ash
(2) lived in shallow water
(4) existed for a geologically short time

48. The table below gives information about the radioactive decay of carbon-14. [Part of the table has been left blank for student use.]

Half-life	Mass of Original C-14 Remaining (grams)	Number of Years
0	1	0
1	$\frac{1}{2}$	5,700
2	$\frac{1}{4}$	11,400
3	$\frac{1}{8}$	17,100
4		
5		
6		

What is the amount of the original carbon-14 remaining after 34,200 years?
(1) $\frac{1}{8}$g
(3) $\frac{1}{32}$g
(2) $\frac{1}{16}$g
(4) $\frac{1}{64}$g

49. According to the *Earth Science Reference Tables*, when did the Jurassic Period end?
(1) 66 million years ago
(3) 163 million years ago
(2) 144 million years ago
(4) 190 million years ago

50. According to the *Earth Science Reference Tables*, much of the surface bedrock of the Adirondack Mountains consists of
(1) gneiss and quartzite
(3) conglomerate and red shale
(2) limestone and sandstone
(4) slate and dolostone

51. According to the *Earth Science Reference Tables*, which geologic event is associated with the Grenville Orogeny?
(1) the initial opening of the Atlantic Ocean
(2) the separation of South America from Africa
(3) the formation of the ancestral Adirondack Mountains
(4) the advance and retreat of the last continental ice sheet

52. In which type of landscape are meandering streams most likely found?
(1) regions of waterfalls
(3) steeply sloping hills
(2) gently sloping plains
(4) V-shaped valleys

53. Which cross section best represents the general bedrock structure of New York State's Allegheny Plateau?

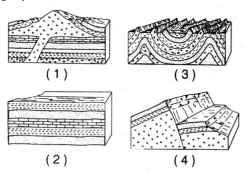

(1)   (3)

(2)   (4)

54. The diagram below represents a geologic cross section.

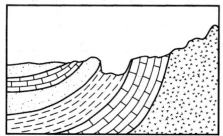

Which rock type appears to have weathered and eroded most?

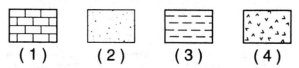

(1)   (2)   (3)   (4)

Note that question 55 has only three choices.

55. If a sample of a radioactive substance is crushed, the half-life of the substance will
   (1) decrease
   (2) increase
   (3) remain the same

PART II

This part consists of ten groups, each containing five questions. Choose seven of these ten groups. Be sure that you answer all five questions in each group chosen. [35]

Group 1

If you choose this group, be sure to answer questions 56–60.

Base your answers to questions 56 through 60 on the *Earth Science Reference Tables,* the topographic map below, and your knowledge of Earth science. Points *A* and *B* represent locations on the map. Elevations are shown in meters.

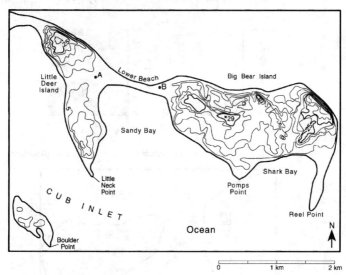

56. What is the approximate distance across Cub Inlet from Little Neck Point to Boulder Point?
    (1) 1.4 km   (3) 2.4 km
    (2) 1.9 km   (4) 2.9 km

57. What could be the highest elevation on Little Deer Island?
    (1) 5 m   (3) 34 m
    (2) 25 m   (4) 39 m

58. Which section of Big Bear Island has the steepest coastline?
    (1) northwest   (3) southwest
    (2) northeast   (4) southeast

59. Which diagram best represents the topographic profile from location A to location B?

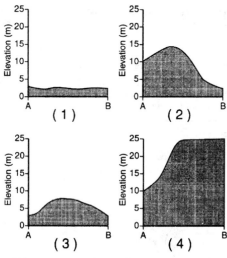

60. Which map would best represent this area if sea level were to rise 10 meters?

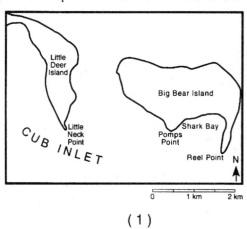

(1)

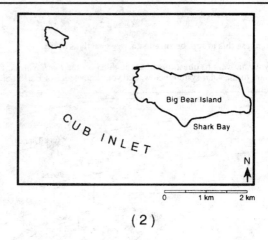

(2)

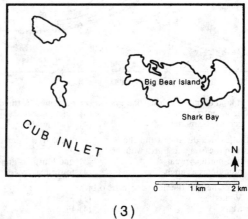

(3)

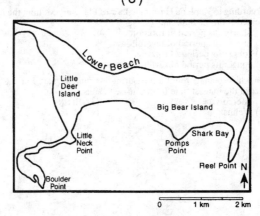

(4)

Group 2

If you choose this group, be sure to answer questions 61–65.

Base your answers to questions 61 through 65 on the *Earth Science Reference Tables,* the diagram of the solar system below, and your knowledge of Earth science.

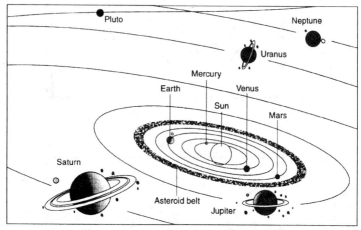

(Not drawn to scale)

61. What kind of model of the solar system is represented by the diagram?
    (1) heliocentric model
    (2) geocentric model
    (3) sidereal model
    (4) lunar model

62. If the Earth's distance from the Sun were doubled, the gravitational attraction between the Sun and Earth would be
    (1) one-ninth as great
    (2) nine times as great
    (3) one-fourth as great
    (4) four times as great

63. Which planet has the most eccentric orbit?
    (1) Venus
    (2) Mars
    (3) Saturn
    (4) Pluto

64. According to Kepler's Harmonic Law of Planetary Motion, the farther a planet is located from the Sun, the
    (1) shorter its period of rotation
    (2) shorter its period of revolution
    (3) longer its period of rotation
    (4) longer its period of revolution

65. On which planet would a measuring instrument placed at the planet's equator record the longest time from sunrise to sunset?
    (1) Mercury
    (2) Venus
    (3) Earth
    (4) Mars

Group 3

If you choose this group, be sure to answer questions 66–70.

Base your answers to questions 66 through 70 on the diagram below and on your knowledge of Earth science. Two identical houses, *A* and *B*, were built in a city in New York State. One house was built on the east side of a factory, and the other house was built on the west side of the factory. Both houses originally had white roofs, but the roof on house *B* has been blackened by factory soot falling on it over the years.

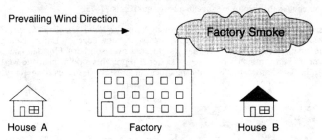

66. Which graph best shows the amount of insolation received at house *A* during the daylight hours?

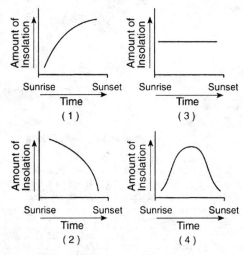

67. On which date would the greatest amount of insolation normally be received at house *A*?
    (1) December 21   (3) June 21
    (2) March 21   (4) September 21

68. Compared to the amount of insolation reflected by the roof of house *A*, the amount of insolation reflected by the roof of house *B* is
    (1) usually less
    (2) usually greater
    (3) always the same
    (4) less in summer and greater in winter

69. Energy is transferred from the Sun to houses *A* and *B* primarily by
    (1) refraction   (3) conduction
    (2) radiation   (4) convection

Note that question 70 has only three choices.

70. As the amount of dust and smoke in the air over house *B* increases, the amount of insolation that reaches house *B* will
(1) decrease
(2) increase
(3) remain the same

Group 4

If you choose this group, be sure to answer questions 71–75.

Base your answers to questions 71 through 75 on the *Earth Science Reference Tables,* the weather map and barogram below, and your knowledge of Earth science. The weather map shows a hurricane that was located over southern Florida. The isobars show air pressure in inches of mercury. Letter *A* represents a point near the west coast of Florida. The barogram shows the recorded air pressure in inches of mercury as the hurricane passed near Miami, Florida.

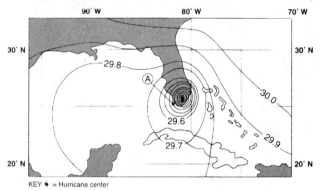

KEY ✷ = Hurricane center

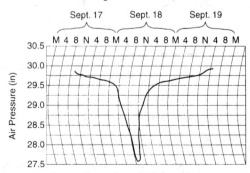

Barogram for Miami, Florida

71. What is the latitude and longitude at the center of the hurricane?
(1) 26° N 81° W   (3) 34° N 81° W
(2) 26° N 89° W   (4) 34° N 89° W

72. What was the lowest air pressure recorded on the barogram as the hurricane passed near Miami?
(1) 27.30 in
(2) 27.60 in
(3) 27.75 in
(4) 28.60 in

73. Which station model best represents the weather conditions at point *A*?

(1) (3)

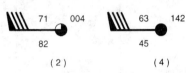

(2) (4)

74. Which type of air mass would most likely be the source of the moisture that causes the strong winds and heavy rain associated with this hurricane?
(1) cP
(2) cT
(3) mP
(4) mT

75. Which map shows the most likely track of this hurricane?

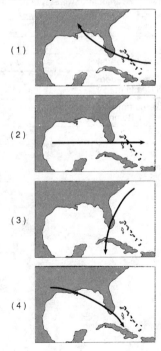

Group 5

If you choose this group, be sure to answer questions 76–80.

Base your answers to questions 76 through 80 on the *Earth Science Reference Tables*, the diagram, data table, and information below, and your knowledge of Earth science.

The diagram below represents part of the laboratory setup for an activity to investigate the effects of particle size on permeability, porosity, and water retention. Three separate tubes were used, each containing 300 milliliters of beads of uniform size. Bead sizes were 4 millimeters, 7 millimeters, and 12 millimeters in diameter, respectively.

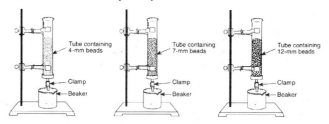

The amount of water added to each tube to cover the beads was determined. The clamp was then removed, the flow of the water was timed, and its volume was measured. Data are shown in the table below. (The amount of water retained on the 7-millimeter beads has been omitted.)

	Particle Size		
	4 mm	7 mm	12 mm
Infiltration Time (seconds)	3.7	3.0	2.4
Amount of Water Needed To Cover All Beads (mL)	147	145	147
Water Recovered from Tube After Clamp Was Removed (mL)	111	123	135
Water Retained on Beads (mL)	36		12

76. Which graph best represents the infiltration times for these three particle sizes?

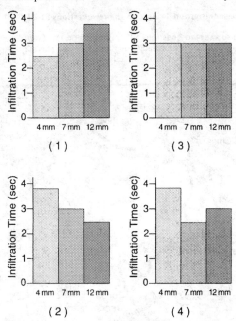

77. What was the total amount of water retained on the 7-millimeter beads after the tubing was unclamped and the water flowed out?
    (1) 8 mL   (3) 22 mL
    (2) 12 mL   (4) 36 mL

78. The data table shows that all three tubes of beads had approximately the same
    (1) porosity   (3) water retention
    (2) permeability time   (4) capillarity

79. Soil composed of which kind of particles would have the longest infiltration time? [Assume that all particles allow some water to pass through.]
    (1) pebbles   (3) silt
    (2) sand   (4) clay

80. Water can infiltrate loose soil when the soil is
    (1) saturated and permeable
    (2) saturated and impermeable
    (3) unsaturated and permeable
    (4) unsaturated and impermeable

Group 6

If you choose this group, be sure to answer questions 81–85.

Base your answers to questions 81 through 85 on the *Earth Science Reference Tables,* the map below, and your knowledge of Earth science. The map shows mid-ocean ridges and trenches in the Pacific Ocean. Specific areas *A, B, C,* and *D* are indicated by shaded rectangles.

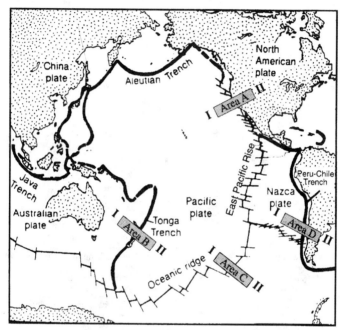

81. Movement of the crustal plates shown in the diagram is most likely caused by
    (1) the revolution of the Earth
    (2) the erosion of the Earth's crust
    (3) shifting of the Earth's magnetic poles
    (4) convection currents in the Earth's mantle

82. The crust at the mid-ocean ridges is composed mainly of
    (1) granite   (3) shale
    (2) basalt   (4) limestone

83. Mid-ocean ridges such as the East Pacific Rise and the Oceanic Ridge are best described as
    (1) mountains containing folded sedimentary rocks
    (2) mountains containing fossils of present-day marine life
    (3) sections of the ocean floor that contain the youngest oceanic crust
    (4) sections of the ocean floor that are the remains of a submerged continent

84. According to the *Earth Science Reference Tables,* which map best shows the direction of movement of the oceanic crustal plates in the vicinity of the East Pacific Rise (ridge)?

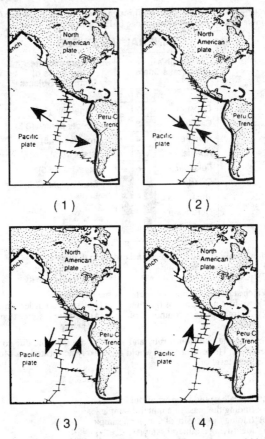

(1) (2)

(3) (4)

85. The cross section below represents an area of the Earth's crust within the map region.

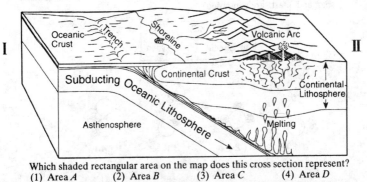

Which shaded rectangular area on the map does this cross section represent?
(1) Area *A* (2) Area *B* (3) Area *C* (4) Area *D*

Group 7

If you choose this group, be sure to answer questions 86–90.

Base your answers to questions 86 through 90 on the *Earth Science Reference Tables*, the data table below, and your knowledge of Earth science.

MINERAL HARDNESS

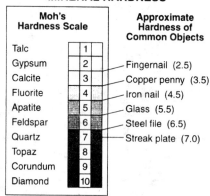

86. Moh's scale arranges minerals according to their relative
    (1) resistance to breaking
    (2) resistance to scratching
    (3) specific heat
    (4) specific gravity

87. Which statement is best supported by the data shown?
    (1) An iron nail contains fluorite.
    (2) A streak plate is composed of quartz.
    (3) Topaz is harder than a steel file.
    (4) Apatite is softer than a copper penny.

88. The durable gemstones ruby and sapphire are valuable due to their color and hardness. These gemstones would most likely be located on Moh's scale at the hardness level of
    (1) 1
    (2) 9
    (3) 3
    (4) 4

89. Moh's scale would be most useful for
    (1) finding the mass of a mineral sample
    (2) finding the density of a mineral sample
    (3) identifying a mineral sample
    (4) counting the number of cleavage surfaces of a mineral sample

90. Which diagram best represents the silicon-oxygen tetrahedron of which talc, feldspar, and quartz are composed?

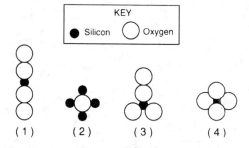

Group 8

If you choose this group, be sure to answer questions 91–95.

Base your answers to questions 91 through 95 on the *Earth Science Reference Tables,* the diagrams below, and your knowledge of Earth science. The diagrams represent two rock outcrops found several miles apart in New York State. Individual rock layers are lettered, and fossils and rock types are indicated.

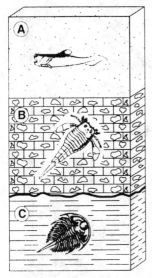 

91. Which processes were directly involved in the formation of these rock layers?
   (1) melting and solidification
   (2) heating and pressure
   (3) compaction and cementation
   (4) conduction and convection

92. An unconformity (buried erosional surface) is represented by the interface between which two layers?
   (1) *A* and *B*
   (2) *B* and *C*
   (3) *D* and *E*
   (4) *E* and *F*

93. In which sequence are the rock layers listed in order from oldest to youngest?
   (1) *F, B, E, D*
   (2) *C, A, F, D*
   (3) *F, E, C, A*
   (4) *C, E, D, A*

94. According to the "Scheme for Sedimentary Rock Identification" in the *Earth Science Reference Tables,* which rock layer was formed mainly from organically formed sediments in seawater?
   (1) *A*
   (2) *B*
   (3) *E*
   (4) *F*

95. Based on the given rock and fossil evidence, which two letters most likely indicate parts of the same layer?
   (1) *A* and *F*
   (2) *B* and *D*
   (3) *C* and *E*
   (4) *D* and *A*

Group 9

If you choose this group, be sure to answer questions 96–100.

Base your answers to questions 96 through 100 on the *Earth Science Reference Tables,* the map and profile below, and your knowledge of Earth science. The map shows part of the canal system of New York State. The profile shows the Erie Canal between Lake Erie and Troy, New York. The numbers in parentheses represent the elevations, in feet, of the canal at the cities indicated.

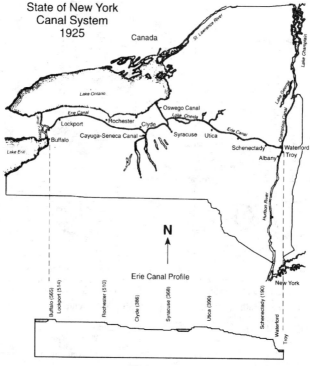

State of New York Canal System 1925

Erie Canal Profile

(Not drawn to scale)

96. Most of the bedrock removed during the digging of the Erie Canal was
    (1) sedimentary
    (2) metamorphic
    (3) intrusive igneous
    (4) extrusive igneous

97. Which river makes up a major portion of the Erie Canal from Syracuse to Schenectady?
    (1) Genesee
    (2) Susquehanna
    (3) Delaware
    (4) Mohawk

98. What is the approximate change in elevation from Buffalo to Clyde?
    (1) 179 ft
    (2) 286 ft
    (3) 515 ft
    (4) 566 ft

99. Most of the Erie Canal was built across which two landscape regions in New York State?
    (1) Erie-Ontario Lowlands and Adirondack Mountains
    (2) Hudson-Mohawk Lowlands and Erie-Ontario Lowlands
    (3) Allegheny Plateau and Hudson-Mohawk Lowlands
    (4) Tug Hill Plateau and the Catskills

100. Which body of water is shown on the canal profile immediately west of Troy, New York?
  (1) Lake Oneida
  (2) Lake George
  (3) St. Lawrence River
  (4) Hudson River

Group 10

If you choose this group, be sure to answer questions 101–105.

Base your answers to questions 101 through 105 on the *Earth Science Reference Tables* and your knowledge of Earth science.

101. Which ocean current provides warm water that moderates the climate of South America?
  (1) Benguela Current
  (2) Brazil Current
  (3) Falkland Current
  (4) Peru Current

102. At which New York State location is the altitude of Polaris closest to 42°?
  (1) Albany
  (2) Rochester
  (3) Slide Mt.
  (4) Mt. Marcy

103. A stream is carrying sediment particles ranging from 0.0004 to 25.6 centimeters. When the stream's velocity decreases from 300 to 100 centimeters per second, the stream will most probably deposit
  (1) silt and clay
  (2) sand and silt
  (3) pebbles and sand
  (4) cobbles and pebbles

104. What is the dewpoint temperature when the dry-bulb temperature is 12°C and the wet-bulb temperature is 7°C?
  (1) 1°C
  (2) −2°C
  (3) −5°C
  (4) 4°C

Note that question 105 has only three choices.

105. Compared to the wavelength of ultraviolet radiation, the wavelength of infrared radiation is
  (1) shorter
  (2) longer
  (3) the same

EARTH SCIENCE

June 18, 1997

PART I

Answer all 55 questions in this part. [55]

Directions (1–55): For *each* statement or question, select the word or expression that, of those given, best completes the statement or answers the question. Some questions may require the use of the *Earth Science Reference Tables*.

1. The circumference of the Earth is about 4.0×10^4 kilometers. This value is equal to
   (1) 400 km (2) 4,000 km (3) 40,000 km (4) 400,000 km

2. The diagram below shows a process of weathering called frost wedging.

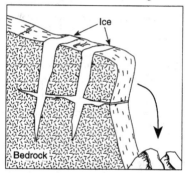

   Frost wedging breaks rocks because as water freezes it increases in
   (1) mass (2) volume (3) density (4) specific heat

3. A student calculates the period of Saturn's revolution to be 31.33 years. What is the student's approximate deviation from the accepted value?
   (1) 1.9% (2) 5.9% (3) 6.3% (4) 19%

4. Which object best represents a true scale model of the shape of the Earth?
   (1) a Ping-Pong ball (3) an egg
   (2) a football (4) a pear

5. Which graph best represents the most common relationship between the amount of air pollution and the distance from an industrial city?

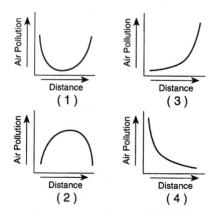

6. Isolines on the topographic map below show elevations above sea level, measured in meters.

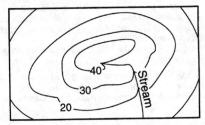

What could be the highest possible elevation represented on this map?
(1) 39 m   (2) 41 m   (3) 45 m   (4) 49 m

7. Oxygen is the most abundant element by volume in the Earth's
(1) inner core   (2) crust   (3) hydrosphere   (4) troposphere

Base your answers to questions 8 and 9 on the diagram below and on your knowledge of Earth science. The diagram represents the path of a planet orbiting a star. Points *A, B, C,* and *D* indicate four orbital positions of the planet.

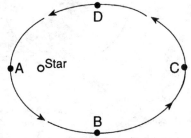

8. When viewed by an observer on the planet, the star has the largest apparent diameter at position
(1) *A*   (2) *B*   (3) *C*   (4) *D*

9. Which graph best represents the gravitational attraction between the star and the planet?

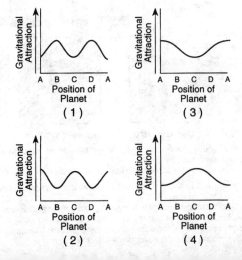

10. Which statement best explains why different phases of the Moon can be observed from the Earth?
    (1) The size of the Earth's shadow falling on the Moon changes.
    (2) The Moon moves into different parts of the Earth's shadow.
    (3) Differing amounts of the Moon's sunlit surface are seen because the Moon revolves around the Sun.
    (4) Differing amounts of the Moon's sunlit surface are seen because the Moon revolves around the Earth.

11. The diagram below represents an activity in which an eye dropper was used to place a drop of water on a spinning globe. Instead of flowing due south toward the target point, the drop followed a curved path and missed the target.

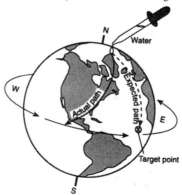

The actual path results from
    (1) the tilt of the globe's axis
    (2) the Coriolis effect
    (3) the globe's revolution
    (4) dynamic equilibrium

Base your answers to questions 12 and 13 on the diagrams below, which show laboratory equipment setups A and B being used to study energy transfer in a classroom laboratory.

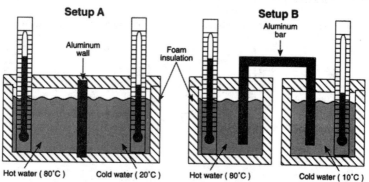

12. In both A and B, most of the heat energy transferred from the hot water to the cold water is transferred by
    (1) convection   (2) conduction   (3) radiation   (4) gravity

13. Which laboratory setup is more efficient at transferring heat energy from the hot water to the cold water?
    (1) A, because less energy is lost to the surrounding environment
    (2) A, because the hot water has a higher temperature
    (3) B, because the aluminum bar is bigger than the aluminum wall
    (4) B, because the cold water has a lower temperature

14. Which diagram best shows how air inside a greenhouse warms as a result of insolation from the Sun?

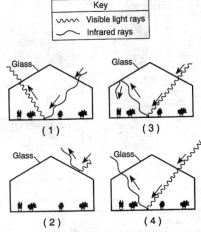

15. Why do the locations of sunrise and sunset vary in a cyclical pattern throughout the year?
    (1) The Earth's orbit around the Sun is an ellipse.
    (2) The Sun's orbit around the Earth is an ellipse.
    (3) The Sun rotates on an inclined axis while revolving around the Earth.
    (4) The Earth rotates on an inclined axis while revolving around the Sun.

16. On a clear April morning near Rochester, New York, which surface will absorb the most insolation per square meter?
    (1) a calm lake   (3) a white-sand beach
    (2) a snowdrift   (4) a freshly plowed farm field

Base your answers to questions 17 and 18 on the weather instrument shown in the diagram below.

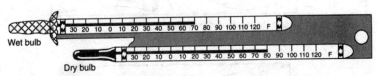

17. What are the equivalent Celsius temperature readings for the Fahrenheit readings shown?
    (1) wet 21°C, dry 27°C   (3) wet 70°C, dry 80°C
    (2) wet 26°C, dry 37°C   (4) wet 158°C, dry 176°C

18. Which weather variables are most easily determined by using this weather instrument and the *Earth Science Reference Tables?*
    (1) air temperature and windspeed   (3) relative humidity and dewpoint
    (2) visibility and wind direction   (4) air pressure and cloud type

19. All of the containers shown below contain the same volume of water and are at room temperature. In a two-day period, from which container will the *least* amount of water evaporate?

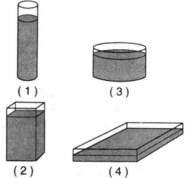

(1)   (3)

(2)   (4)

20. Which statement best explains why precipitation occurs at the frontal surfaces between air masses?
    (1) Warm, moist air sinks when it meets cold, dry air.
    (2) Warm, moist air rises when it meets cold, dry air.
    (3) Cold fronts move faster than warm fronts.
    (4) Cold fronts move slower than warm fronts.

21. By which process does water vapor leave the atmosphere and form dew?
    (1) condensation   (2) transpiration   (3) convection   (4) precipitation

22. The diagram below represents a section of a weather map showing high- and low-pressure systems. The lines represent isobars.

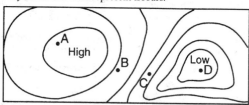

At which point is the windspeed greatest?
(1) A   (2) B   (3) C   (4) D

23. A low-pressure system is shown on the weather map below.

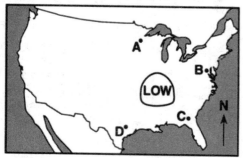

Toward which point will the low-pressure system move if it follows a typical storm track?
(1) A   (2) B   (3) C   (4) D

24. The diagram below shows equal volumes of loosely packed sand, clay, and small pebbles placed in identical funnels. The soils are dry, and the beakers are empty.

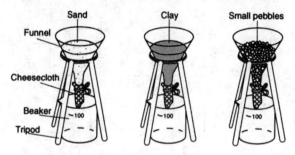

A 100-milliliter sample of water is poured into each funnel at the same time and allowed to seep through for 15 minutes. Which diagram best shows the amount of water that passes through each funnel into the beakers?

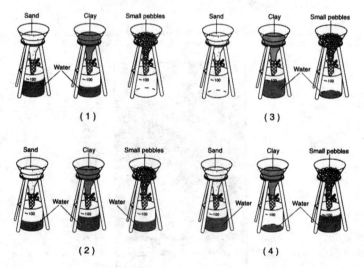

25. The upward movement of water through tiny spaces in soil or rock is called
    (1) water retention
    (2) capillary
    (3) porosity
    (4) permeability

26. A deposit of rock particles that are angular, scratched, and unsorted has most likely been transported and deposited by
    (1) ocean waves
    (2) running water
    (3) a glacier
    (4) wind

27. When the soil is saturated in a gently sloping area, any additional rainfall in the area will most likely
    (1) become ground water
    (2) become surface runoff
    (3) cause a moisture deficit
    (4) cause a higher potential evapotranspiration

28. Which graph best represents conditions in an area with a moisture deficit?
[Key: E_p = potential evapotranspiration; P = precipitation; St = storage]

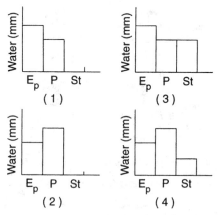

29. The diagram below shows a residual soil profile formed in an area of granite bedrock. Four different soil horizons, A, B, C, and D, are shown.

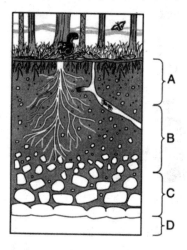

Which soil horizon contains the greatest amount of material formed by biological activity?

(1) A   (2) B   (3) C   (4) D

Base your answers to questions 30 and 31 on the diagram below. The diagram shows points *A, B, C,* and *D,* on a meandering stream.

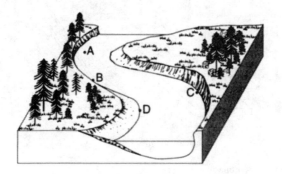

30. Which material is most likely to be transported in suspension during periods of slowest stream velocity?
    (1) gravel
    (2) sand
    (3) silt
    (4) clay

31. At which point is the amount of deposition more than the amount of erosion?
    (1) *A*
    (2) *B*
    (3) *C*
    (4) *D*

32. Which rock is usually composed of several different minerals?
    (1) gneiss
    (2) quartzite
    (3) rock gypsum
    (4) chemical limestone

33. Which granite sample most likely formed from magma that cooled and solidified at the slowest rate?

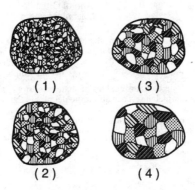

34. The diagram below shows the results of one test for mineral identification.

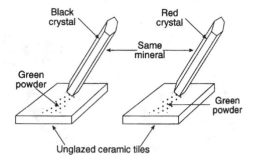

Which mineral property is being tested?
(1) density   (3) streak
(2) fracture   (4) luster

35. Which statement about the formation of a rock is best supported by geologic evidence?
(1) Magma must be weathered before it can change to metamorphic rock.
(2) Sediment must be compacted and cemented before it can change to sedimentary rock.
(3) Sedimentary rock must melt before it can change to metamorphic rock.
(4) Metamorphic rock must melt before it can change to sedimentary rock.

36. The scale below shows the age of the sea-floor crust in relation to its distance from the Mid-Atlantic Ridge.

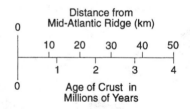

Crust that originally formed at the Mid-Atlantic Ridge is now 37 kilometers from the ridge. Approximately how long ago did this crust form?
(1) 1.8 million years ago   (3) 3.0 million years ago
(2) 2.0 million years ago   (4) 4.5 million years ago

37. A conglomerate contains pebbles of shale, sandstone, and granite. Based on this information, which inference about the pebbles in the conglomerate is most accurate?
(1) They were eroded by slow-moving water.
(2) They came from other conglomerates.
(3) They are all the same age.
(4) They had various origins.

38. The diagrams below show demonstrations that represent the behavior of two seismic waves, *A* and *B*.

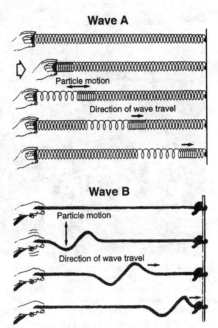

Which statement concerning the demonstrated waves is correct?
(1) Wave *A* represents a compressional wave, and wave *B* represents a shear wave.
(2) Wave *A* represents a shear wave, and wave *B* represents a compressional wave.
(3) Wave *A* represents compressional waves in the crust, and wave *B* represents compressional waves in the mantle.
(4) Wave *A* represents shear waves in the crust, and wave *B* represents shear waves in the mantle.

39. The cartoon below presents a humorous view of Earth science.

The cartoon character on the right realizes that the sand castle will eventually be
(1) preserved as fossil evidence
(2) deformed during metamorphic change
(3) removed by agents of erosion
(4) compacted into solid bedrock

40. The diagram below represents a cross section of a portion of the Earth's lithosphere.

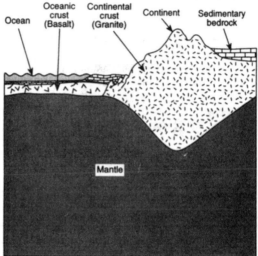

Which statement about the Earth's crust is best supported by the diagram?
(1) The crust is thicker than the mantle.
(2) The continental crust is thicker than the oceanic crust.
(3) The crust is composed primarily of sedimentary rock.
(4) The crust is composed of denser rock than the mantle.

41. The epicenter of an earthquake is 6,000 kilometers from an observation point. What is the difference in travel time for the P-waves and S-waves?
(1) 7 min 35 sec   (2) 9 min 20 sec   (3) 13 min 10 sec   (4) 17 min 00 sec

42. At a depth of 2,000 kilometers, the temperature of the stiffer mantle is inferred to be
(1) 6,500°C   (2) 4,200°C   (3) 3,500°C   (4) 1,500°C

43. Hot springs on the ocean floor near the mid-ocean ridges provide evidence that
(1) climate change has melted huge glaciers
(2) marine fossils have been up lifted to high elevations
(3) meteor craters are found beneath the oceans
(4) convection currents exist in the asthenosphere

44. The diagram below represents the skull of a saber-toothed tiger that died 30,000 years ago.

The age of the skull could be determined most accurately by using
(1) uranium-238   (2) rubidium-87   (3) potassium-40   (4) carbon-14

45. A geologist collected the fossils shown below from locations in New York State.

Which sequence correctly shows the fossils from oldest to youngest?

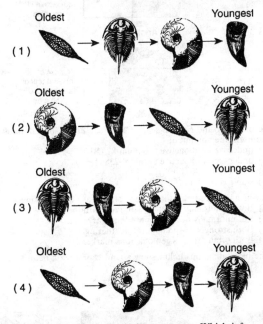

46. Present-day corals live in warm, tropical ocean water. Which inference is best supported by the discovery of Ordovician-age corals in the surface bedrock of western New York State?
    (1) Western New York State was covered by a warm, shallow sea during Ordovician time.
    (2) Ordovician-age corals lived in the forests of western New York State.
    (3) Ordovician-age corals were transported to western New York State by cold, freshwater streams.
    (4) Western New York State was covered by a continental ice sheet that created coral fossils of Ordovician time.

47. The diagram below represents the radioactive decay of uranium-238.

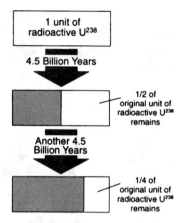

According to the *Earth Science Reference Tables*, shaded areas on the diagram represent the amount of
(1) undecayed radioactive rubidium-87 (Rb^{87})
(2) undecayed radioactive uranium-238 (U^{238})
(3) stable lead-206 (Pb^{206})
(4) stable carbon-14 (C^{14})

48. In which way are index fossils and volcanic ash deposits similar?
(1) Both can usually be dated with radiocarbon.
(2) Both normally occur in nonsedimentary rocks.
(3) Both strongly resist chemical weathering.
(4) Both often serve as geologic time markers.

49. The block diagrams below show a landscape region before and after uplift and erosion.

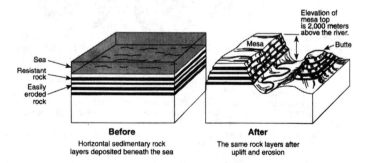

The landscape shown in the "after" diagram is best classified as a
(1) folded mountain   (3) plains region
(2) plateau region   (4) volcanic dome

50. Which statement correctly describes an age relationship in the geologic cross section below?

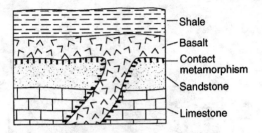

(1) The sandstone is younger than the basalt.
(2) The shale is younger than the basalt.
(3) The limestone is younger than the shale.
(4) The limestone is younger than the sandstone.

51. The landscape around Old Forge, New York, consists of
    (1) flat, low-lying volcanic rock
    (2) hills of horizontal sedimentary rock layers
    (3) mountains of metamorphic rock
    (4) eroded, tilted sedimentary rock

52. The diagrams below represent geologic cross sections from two widely separated regions.

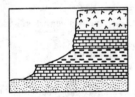 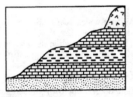

The layers of rock appear very similar, but the hillslopes and shapes are different. These differences are most likely the result of
    (1) volcanic eruptions
    (2) earthquake activity
    (3) soil formation
    (4) climate variations

53. Which bedrock characteristics most influence landscape development?
    (1) composition and structure
    (2) structure and age
    (3) age and color
    (4) color and composition

54. The map below shows the location of an ancient sea, which evaporated to form the Silurian-age deposits of rock salt and rock gypsum now found in some New York State bedrock.

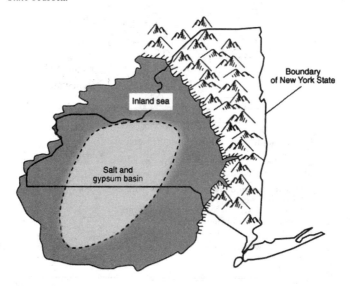

Within which two landscape regions are these large rock salt and rock gypsum deposits found?
(1) Hudson Highlands and Taconic Mountains
(2) Tug Hill Plateau and Adirondack Mountains
(3) Erie-Ontario Lowlands and Allegheny Plateau
(4) Catskills and Hudson-Mohawk Lowlands

55. The diagram below represents a cross section of the bedrock and land surface in part of Tennessee. The dotted lines indicate missing rock layers.

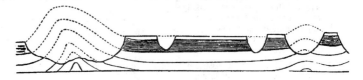

Which statement is best supported by the diagram?
(1) Rocks are weathered and eroded evenly.
(2) Folded rocks are more easily weathered and eroded.
(3) Deposits of sediments provide evidence of erosion.
(4) Climate differences affect the amount of erosion.

PART II

This part consists of ten groups, each containing five questions. Choose seven of these ten groups. Be sure that you answer all five questions in each group chosen. [35]

Group 1

If you choose this group, be sure to answer questions 56–60.

Base your answers to questions 56 through 60 on the table below and on your knowledge of Earth science. The table shows data for a student's collection of rock samples *A* through *I*, which are classified into groups *X*, *Y*, and *Z*. For each rock sample, the student recorded mass, volume, density, and a brief description. The density for rock *D* has been left blank.

Rock Collection

Group	Rock	Mass (g)	Volume (cm³)	Density (g/cm³)	Description
X	A	82.9	34.4	2.41	Grey, smooth, rounded
	B	114.2	42.6	2.68	Brown, smooth, rounded
	C	144.7	63.2	2.29	Black, smooth, rounded
Y	D	159.4	59.7		Black and grey crystals, angular
	E	87.7	33.1	2.65	Clear and pink crystals, angular
	F	59.6	21.0	2.84	White, grey, and black crystals; angular
Z	G	201.1	68.4	2.94	Grey, shiny, flat
	H	85.1	39.8	2.14	Brown, sandy feel, flat
	I	110.2	47.3	2.33	Dark grey, flaky, flat

56. The student's classification system is based on
    (1) density (2) shape (3) color (4) mass

57. The approximate density of rock sample *D* is
    (1) 2.67 g/cm³ (2) 2.75 g/cm³ (3) 3.32 g/cm³ (4) 3.75 g/cm³

58. Which statement is an inference rather than an observation?
    (1) Rock *H* is flat.
    (2) Rock *B* has been rounded by stream action.
    (3) Rock *E* has a volume of 33.1 cm³.
    (4) Rock *G* is the same color as rock *I*.

59. To obtain the data recorded in the column labeled "Description," the student used
    (1) a triple-beam balance   (3) a calculator
    (2) an overflow can   (4) her senses

Note that question 60 has only three choices.

60. The student broke rock *G* into two pieces. Compared to the density of the original rock, the density of one piece would most likely be
    (1) less
    (2) greater
    (3) the same

Group 2

If you choose this group, be sure to answer questions 61–65.

Base your answers to questions 61 through 65 on the topographic map below and on your knowledge of Earth science. Points *A*, *B*, *C*, *D*, *E*, *F*, *X*, and *Y* are locations on the map. Elevation is measured in feet.

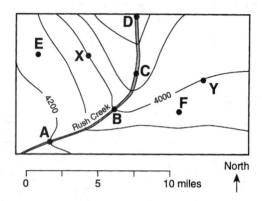

61. What is the contour interval used on this map?
    (1) 20 ft   (2) 50 ft   (3) 100 ft   (4) 200 ft

62. Which locations have the greatest difference in elevation?
    (1) *A* and *D*   (2) *B* and *X*   (3) *C* and *F*   (4) *E* and *Y*

63. Between points *C* and *D*, Rush Creek flows toward the
    (1) north   (2) south   (3) east   (4) west

64. The gradient between points *A* and *B* is closest to
    (1) 20 ft/mi   (2) 40 ft/mi   (3) 80 ft/mi   (4) 200 ft/mi

65. Which diagram best represents the profile along a straight line between points *X* and *Y*?

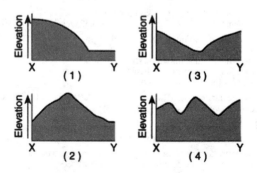

Group 3

If you choose this group, be sure to answer questions 66–70.

Base your answers to questions 66 through 70 on the *Earth Science Reference Tables*, the map below, and your knowledge of Earth science. The map represents a view of the Earth looking down from above the North Pole (N.P.), showing the Earth's 24 standard time zones. The Sun's rays are striking the Earth from the right. Points *A*, *B*, *C*, and *D* are locations on the Earth's surface.

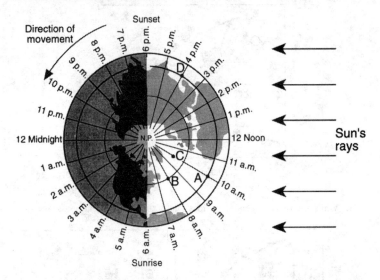

66. The timekeeping system shown in the diagram is based on the
    (1) Sun's revolution   (3) Earth's revolution
    (2) Sun's rotation   (4) Earth's rotation

67. Which date could this diagram represent?
    (1) January 21   (3) June 21
    (2) March 21   (4) August 21

68. Which two points have the same longtitude?
    (1) *A* and *C*   (2) *B* and *C*   (3) *A* and *D*   (4) *B* and *D*

69. Areas within a time zone generally keep the same standard clock time. In degrees of longitude, approximately how wide is one standard time zone?
    (1) 7½°   (2) 15°   (3) 23½°   (4) 30°

70. At which position would the altitude of the North Star (Polaris) be greatest?
    (1) *A*   (2) *B*   (3) *C*   (4) *D*

Group 4

If you choose this group, be sure to answer questions 71–75.

Base your answers to questions 71 through 75 on the *Earth Science Reference Tables,* the weather map below, and your knowledge of Earth science. The map shows a low-pressure system over the eastern part of the United States. Weather data is given for cities *A* through *I*. The temperature at city *E* has been left blank.

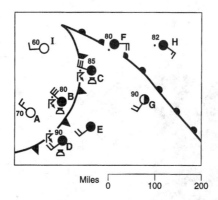

71. Which map correctly shows the locations of the cP and mT air-mass labels?

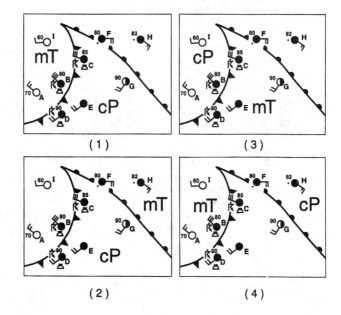

72. The symbol ⛁ represents a cumulonimbus cloud. What is the most probable explanation for the absence of this cloud symbol at city A?
    (1) City A's atmosphere lacks the necessary moisture.
    (2) City A is located ahead of the cold front.
    (3) Cumulonimbus clouds form only at temperatures higher than 70°F.
    (4) Cumulonimbus clouds form only when a location has southwesterly winds.

73. What is the most probable temperature for city E?
    (1) 60°F   (2) 70°F   (3) 75°F   (4) 88°F

74. Which city is *least* likely to have precipitation in the next few hours?
    (1) A   (2) F   (3) C   (4) H

75. Which map correctly shows arrows indicating the surface wind pattern?

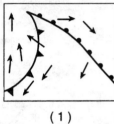

(1)

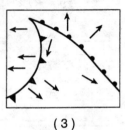

(3)

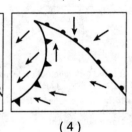

(2)   (4)

Group 5

If you choose this group, be sure to answer questions 76–80.

Base your answers to questions 76 through 80 on the *Earth Science Reference Tables,* the diagram below, and your knowledge of Earth science. The diagram shows the latitude zones of the Earth.

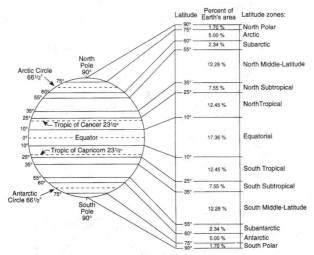

76. What is the total number of degrees of latitude covered by the Equatorial zone?
    (1) 0°   (2) 10°   (3) 17°   (4) 20°

77. In which latitude zone is New York State located?
    (1) North Middle-Latitude   (3) North Tropical
    (2) North Subtropical   (4) North Polar

78. Which graph best represents the relationship between average yearly temperatures and latitude?

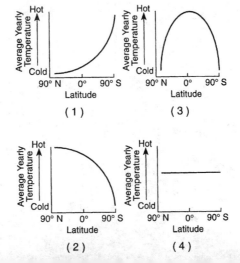

79. Which zone receives the greatest intensity of sunlight on June 21?
    (1) North Tropical   (3) South Tropical
    (2) Equatorial   (4) North Polar

80. The locations of the Tropic of Cancer and the Tropic of Capricorn are set at 23½° from the Equator because the
    (1) Earth is slightly bulged at the equatorial region
    (2) direct rays of the Sun move between these latitudes
    (3) Arctic Circle and the Antarctic Circle are 23½° from the poles
    (4) center of the Earth's gravitational field is located 23½° from the Equator

Group 6

If you choose this group, be sure to answer questions 81–85.

Base your answers to questions 81 through 85 on the *Earth Science Reference Tables,* the maps and cross section below, and your knowledge of Earth science. The maps show the stages in the growth of a stream delta. Point X represents a location in the stream channel. The side view of a stream shows rock particles transported in the stream at a point close to its source.

Maps: Stages in Growth of a Stream Delta

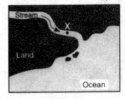
Early stage

Middle stage

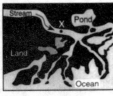
Late stage

Side View of a Stream

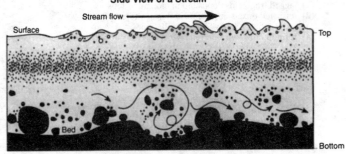

81. The rock materials transported in the stream are most likely transported by which methods?
    (1) in solution, only
    (2) in suspension, only
    (3) in solution and in suspension, only
    (4) in solution, in suspension, and by rolling

82. The velocity of the stream at location X is controlled primarily by the
    (1) amount of sediment carried at location X
    (2) distance from location X to the stream source
    (3) slope of the stream at location X
    (4) temperature of the stream at location X

83. Which graph best illustrates the effect that changes in stream discharge have on stream velocity at location *X*?

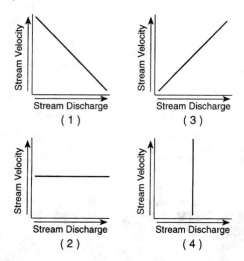

84. A decrease in the velocity of the stream at location *X* will usually cause an increase in
    (1) downcutting by the stream
    (2) deposition within the stream channel
    (3) the size of the particles carried by the stream
    (4) the amount of material carried by the stream

85. Which characteristics are most likely shown by the sediments in the delta?
    (1) jagged fragments deposited in elongated hills
    (2) unsorted mixed sizes deposited in scattered piles
    (3) large cobbles deposited in parallel lines
    (4) round grains deposited in layers

Group 7

If you choose this group, be sure to answer questions 86–90.

Base your answers to questions 86 through 90 on the *Earth Science Reference Tables,* the information below, and your knowledge of Earth science. The map shows surface geology of a portion of the Schoharie Valley in New York State. Patterns and letters are used to indicate bedrock of different ages. The Schoharie Valley contains mostly horizontal rock structure in which overturning has not occurred. The table provides information about the rocks shown on the map.

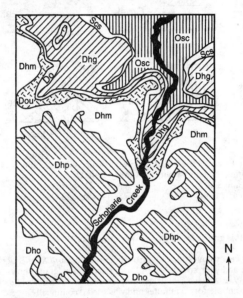

Age of the Rock	Symbol	Composition of the Rock
Middle Devonian	Dho	Shale, sandstone
	Dhp	Shale, siltstone, sandstone
	Dhm	Shale, sandstone, limestone
Early Devonian	Dou	Limestone, shale, siltstone
	Do	Sandstone, limestone
	Dhg	Limestone, dolostone
Late Silurian	Scs	Limestone, shale, dolostone
Middle–Late Ordovician	Osc	Sandstone, siltstone, shale

86. What is the age of the surface bedrock in this portion of the Schoharie Valley?
    (1) 320–374 million years
    (2) 374–478 million years
    (3) 478–505 million years
    (4) 505–540 million years

87. The surface bedrock of this portion of the Schoharie Valley is composed mainly of
    (1) metamorphic rock
    (2) sedimentary rock
    (3) extrusive igneous rock
    (4) intrusive igneous rock

88. Which fossil could be found in the surface bedrock of this portion of the Schoharie Valley?
    (1) figlike leaf
    (2) mastodont
    (3) brachiopod
    (4) coelophysis

89. Which cross section represents a possible arrangement of rock units in a cliff along this portion of Schoharie Creek?

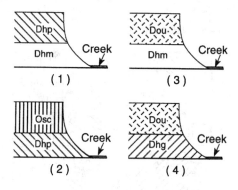

90. Based on the age of the bedrock, between which two rock units is an unconformity representing the longest period of time located?
    (1) Dhg and Scs
    (2) Dhm and Dou
    (3) Do and Dhg
    (4) Scs and Osc

Group 8

If you choose this group, be sure to answer questions 91–95.

Base your answers to questions 91 through 95 on the *Earth Science Reference Tables,* the graph below, and your knowledge of Earth science. The graph shows the results of a laboratory activity in which a 200-gram sample of ice at −50°C was heated in an open beaker at a uniform rate for 70 minutes and was stirred continually.

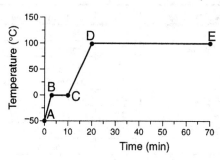

91. What was the temperature of the water 17 minutes after the heating began?
    (1) 0°C
    (2) 12°C
    (3) 75°C
    (4) 100°C

92. Which change occurred between point *A* and point *B*?
    (1) Ice melted.
    (2) Ice warmed.
    (3) Water froze.
    (4) Water condensed.

93. What was the total amount of energy absorbed by the sample during the time between points B and C on the graph?
   (1) 200 calories
   (2) 800 calories
   (3) 10,800 calories
   (4) 16,000 calories

94. During which time interval was the greatest amount of energy added to the water?
   (1) A to B (2) B to C (3) C to D (4) D to E

95. Which change could shorten the time needed to melt the ice completely?
   (1) using colder ice
   (2) stirring the sample more slowly
   (3) reducing the initial sample to 100 grams of ice
   (4) reducing the number of temperature readings taken

Group 9

If you choose this group, be sure to answer questions 96–100.

Base your answers to questions 96 through 100 on the *Earth Science Reference Tables*, the map below, and your knowledge of Earth science. The map shows many of the major faults and fractures in the surface bedrock of New York State.

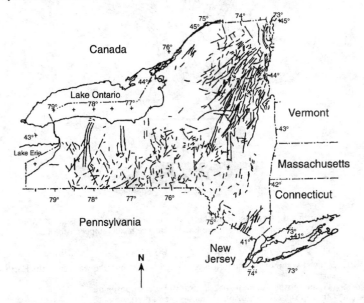

96. In which landscape region do the faults and fractures appear most concentrated?
   (1) Adirondack Mountains
   (2) the Catskills
   (3) Atlantic Costal Plain
   (4) St. Lawrence Lowlands

97. Most of the faults found along the New York–New Jersey border lie in which direction?
   (1) north to south
   (2) east to west
   (3) northwest to southwest
   (4) northeast to southwest

98. The faults and fractures in the bedrock have the greatest effect on the locations and patterns of
   (1) streams
   (2) residual soils
   (3) precipitation
   (4) temperature zones

99. The surface of Long Island shows no visible faults and fractures because the surface is
    (1) a flat plain
    (2) old bedrock
    (3) primarily composed of unconsolidated sediments
    (4) extensively eroded by ocean waves

100. A large earthquake associated with one of these faults occurred at 45° N 75° W on September 5, 1994. Which location in New York State was closest to the epicenter of the earthquake?
    (1) Buffalo
    (2) Massena
    (3) Albany
    (4) New York City

Group 10

If you choose this group, be sure to answer questions 101–105.

101. The diagram below shows air rising from the Earth's surface to form a thunderstorm cloud.

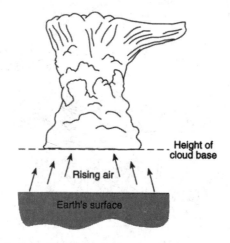

According to the Lapse Rate chart, what is the height of the base of the thunderstorm cloud when the air at the Earth's surface has a temperature of 30°C and a dewpoint of 22°C?
    (1) 1.0 km   (2) 1.5 km   (3) 3.0 km   (4) 0.7 km

102. Which list shows atmospheric layers in the correct order upward from the Earth's surface?
    (1) thermosphere, mesosphere, stratosphere, troposphere
    (2) troposphere, stratosphere, mesosphere, thermosphere
    (3) stratosphere, mesosphere, troposphere, thermosphere
    (4) thermosphere, troposphere, mesosphere, stratosphere

103. Which planet takes longer for one spin on its axis than for one orbit around the Sun?
    (1) Mercury
    (2) Venus
    (3) Earth
    (4) Mars

104. What is the approximate minimum water velocity needed to maintain movement of a sediment particle with a diameter of 5.0 centimeters?
    (1) 75 cm/sec
    (2) 100 cm/sec
    (3) 150 cm/sec
    (4) 200 cm/sec

105. The diagram below illustrates Eratosthenes' method of finding the circumference of a planet. At noon, when a vertical stick at the Equator casts no shadow, a vertical stick 2,500 kilometers away casts a shadow and makes an angle of 40° with the rays of the Sun as shown.

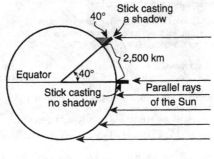

(Not drawn to scale)

What is the circumference of this planet?
(1) 2,500 km
(2) 20,000 km
(3) 22,500 km
(4) 45,000

EARTH SCIENCE

June 18, 1998

PART I

Answer all 55 questions in this part. [55]

Directions (1–55): For *each* statement or question, select the word or expression that, of those given, best completes the statement or answers the question. Some questions may require the use of the *Earth Science Reference Tables.*

1. A student examined a patch of mud and recorded several statements about footprints in the mud. Which statement is most likely an inference?
   (1) There are five footprints in the mud.
   (2) The depth of the deepest footprint is 3 centimeters.
   (3) The footprints were made by a dog.
   (4) The footprints are oriented in an east-west direction.

2. Which statement best explains why water in a glass becomes colder when ice cubes are added?
   (1) The water changes into ice.
   (2) Heat flows from the water to the ice cubes.
   (3) Water is less dense than ice.
   (4) Ice has a higher specific heat than water.

3. The diagram below represents the route of a ship traveling from New York City to Miami, Florida. Each night, a passenger on the ship observes Polaris.

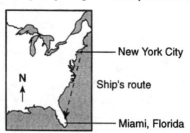

   Which statement best describes the observed changes in the altitude of Polaris made by the passenger during the voyage?
   (1) Each night the altitude decreases in the northern sky.
   (2) Each night the altitude decreases in the southern sky
   (3) Each night the altitude increases in the northern sky.
   (4) Each night the altitude increases in the southern sky.

4. The diagrams below represent photographs of a large sailboat taken through a telescope over time as the boat sailed away from shore out to sea. Each diagram shows the magnification of the lenses and the time of day.

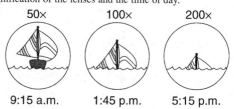

   Which statement best explains the apparent sinking of this sailboat?
   (1) The sailboat is moving around the curved surface of Earth.
   (2) The sailboat appears smaller as it moves farther away.
   (3) The change in density of the atmosphere is causing refraction of light rays.
   (4) The tide is causing an increase in the depth of the ocean.

5. Which observation is a direct result of the $23\frac{1}{2}°$ tilt of Earth's axis as Earth orbits the Sun?
   (1) Locations on Earth's Equator receive 12 hours of daylight every day.
   (2) The apparent diameter of the Sun shows predictable changes in size.
   (3) A Foucault pendulum shows predictable shifts in its direction of swing.
   (4) Winter occurs in the Southern Hemisphere at the same time that summer occurs in the Northern Hemisphere.

6. The diagram below shows spheres associated with Earth.

   Which spheres are zones of Earth's atmosphere?
   (1) lithosphere, hydrosphere, and troposphere
   (2) stratosphere, mesosphere, and thermosphere
   (3) asthenosphere, lithosphere, and hydrosphere
   (4) hydrosphere, troposphere, and stratosphere

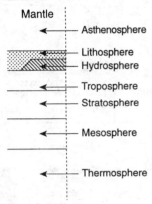

(Not drawn to scale)

7. An observer on Earth measured the apparent diameter of the Sun over a period of 2 years. Which graph best represents the Sun's apparent diameter during the 2 years?

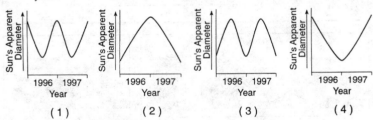

8. Which diagram best represents a portion of the heliocentric model of the solar system? [S = Sun, E = Earth, and M = Moon]

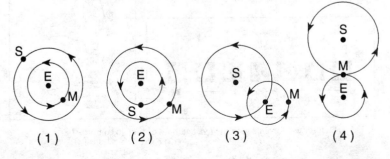

9. The diagram below represents a planet revolving in an elliptical orbit around a star.

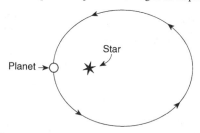

As the planet makes one complete revolution around the star; starting at the position shown, the gravitational attraction between the star and the planet will
(1) decrease, then increase
(2) increase, then decrease
(3) continually decrease
(4) remain the same

10. The diagrams below represent four series of events over the passage of time.

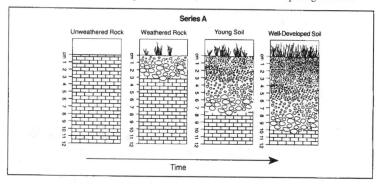

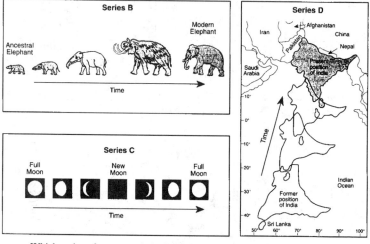

Which series of events took the *least* amount of time to complete?
(1) A   (2) B   (3) C   (4) D

11. The graph below shows the relative amount of radiation energy gained and lost by Earth's surface at all latitudes in July and January.

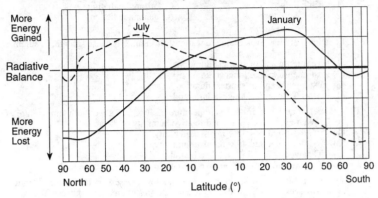

Which statement best explains the differences between the July and January values shown on the graph?
(1) The ozone layer changes position.
(2) The temperature of Earth remains constant.
(3) The position of maximum insolation on Earth's surface changes.
(4) The location of heat flow from Earth's interior changes.

12. Locations *A* and *B* are 600 kilometers apart on the equator of a planet's spherical moon. When the Sun is directly over point *B*, a shadow is cast by a pole at point *A* as shown below.

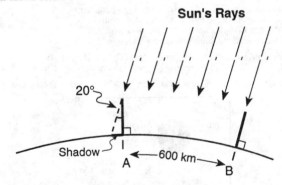

According to Eratosthenes' method, the circumference of the moon would be calculated as
(1) 2,400 km   (2) 3,300 km   (3) 5,400 km   (4) 10,800 km

13. Friction occurring at an interface always produces a
(1) transformation of energy
(2) form of pollution
(3) chemical change
(4) phase change

14. Pieces of lead, copper, iron, and granite, each having a mass of 1 kilogram and a temperature of 100°C, were removed from a container of boiling water and allowed to cool under identical conditions. Which piece most likely cooled to room temperature first?
(1) copper   (2) lead   (3) iron   (4) granite

15. Earth's surface air temperatures change less during cloudy nights than during clear nights because clouds reflect and water vapor absorbs
(1) visible light
(2) ultraviolet light
(3) infrared radiation
(4) gamma radiation

16. Locations in New York State are warmest in summer because sunlight in summer is
(1) least intense and of shortest duration
(2) least intense and of longest duration
(3) most intense and of shortest duration
(4) most intense and of longest duration

17. A parcel of air has a dry-bulb temperature of 16°C and a wet-bulb temperature of 10°C. What are the dewpoint and relative humidity of the air?
(1) –5°C dewpoint and 33% relative humidity
(2) –5°C dewpoint and 45% relative humidity
(3) 4°C dewpoint and 33% relative humidity
(4) 4°C dewpoint and 45% relative humidity

18. Winds are blowing from high-pressure to low-pressure systems over identical ocean surfaces. Which diagram represents the area of greatest windspeed? [Arrows indicate wind direction.]

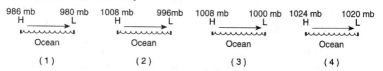

19. Under which set of atmospheric conditions does water usually evaporate at the fastest rate?
(1) warm temperatures, calm winds, and high humidity
(2) warm temperatures, high winds, and low humidity
(3) cold temperatures, calm winds, and low humidity
(4) cold temperatures, high winds, and high humidity

20. A storm system centered over Elmira, New York, will most often track toward
(1) Albany
(2) Jamestown
(3) Rochester
(4) New York City

21. A weather station model for a location in New York State is shown below.

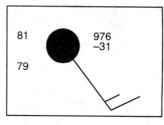

The air mass over this location is best described as
(1) cold with low humidity and high air pressure
(2) cold with high humidity and low air pressure
(3) warm with high humidity and low air pressure
(4) warm with low humidity and high air pressure

22. The diagram below shows a cross section of a cumulus cloud. Line *AB* indicates the base of the cloud.

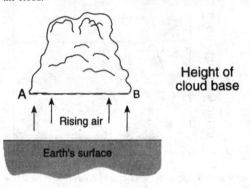

Which graph best represents the temperature measured along line AB?

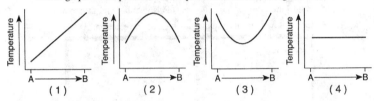

23. The diagram below shows a cross-sectional view of rain falling on a farm field and then moving to the water table.

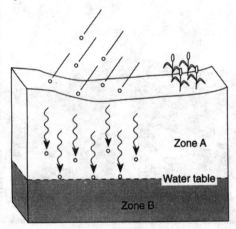

Which word best describes the movement of the rainwater through Zone *A*?
(1) runoff   (2) saturation   (3) infiltration   (4) precipitation

24. The graph below shows the discharge rate of a stream during a 1-year period.

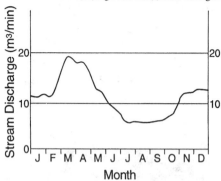

During which time span did a water deficit most likely exist in the water budget for the surrounding area?
(1) January and February
(2) May and June
(3) August and September
(4) November and December

25. The cartoon below presents a humorous look at wave action.

"Here comes another big one, Roy, and here—we—gooooooowheeeeeeeooo!"

The ocean waves that are providing enjoyment for Roy's companion are the result of the
(1) interaction of the hydrosphere with the moving atmosphere
(2) interaction of the lithosphere with the moving troposphere
(3) absorption of short-wave radiation in the stratosphere
(4) absorption of energy in the asthenosphere

26. Which conditions produce the most surface water runoff?
    (1) steep slope, heavy rain, and frozen ground
    (2) steep slope, gentle rain, and unfrozen ground
    (3) gentle slope, heavy rain, and frozen ground
    (4) gentle slope, gentle rain, and unfrozen ground

27. The maps below show changes occurring around a small New York State lake over a 30-year period.

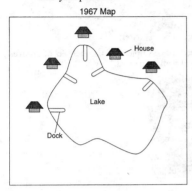

1967 Map

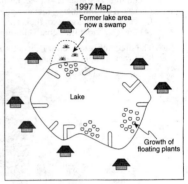

1997 Map

Which graph best shows the probable changes in the quality of ground water and lake water in this region as the changes indicated in the maps took place from 1967 to 1997?

Key
------------ Lake water
———— Ground water

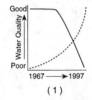

(1)

(2)

(3)

(4)

28. The diagram at the right is a map view of a stream flowing through an area of loose sediments. Arrows show the location of the strongest current.

Which stream profile best represents the cross section from A to A'?

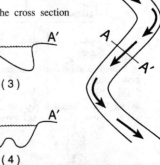

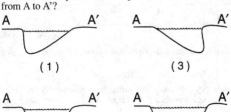

(1)   (3)

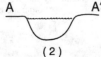

(2)   (4)

29. What occurs when a rock is crushed into a pile of fragments?
    (1) The total surface area decreases and chemical composition changes.
    (2) The total surface area decreases and chemical composition remains the same.
    (3) The total surface area increases and chemical composition changes.
    (4) The total surface area increases and chemical composition remains the same.

30. In which group do the rocks usually have the mineral quartz as part of their composition?
    (1) granite, rhyolite, sandstone, hornfels
    (2) shale, scoria, gneiss, metaconglomerate
    (3) conglomerate, gabbro, rock salt, schist
    (4) breccia, fossil limestone, bituminous coal, siltstone

31. The diagram below shows a hand-sized rock sample with parallel sets of grooves. This rock sample was found in a gravel bank in central New York State.

The grooves were most likely caused by
    (1) stream erosion   (3) a landslide
    (2) wind erosion   (4) glacial erosion

32. The graph below shows the general pattern of erosion and deposition for a small tributary stream. Points *A*, *B*, *C*, and *D* represent locations along the stream.

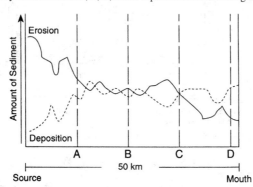

At which location is the erosional-depositional system of the stream in dynamic equilibrium?
    (1) *A*   (2) *B*   (3) *C*   (4) *D*

33. Which kind of sedimentary rock may be formed both chemically and organically?
    (1) limestone   (2) rock gypsum   (3) rock salt   (4) bituminous coal

34. The cross section below represents the transport of sediments by a glacier.

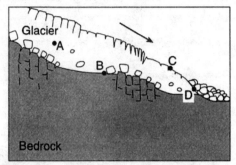

At which location is deposition most likely the dominant process?
(1) A   (2) B   (3) C   (4) D

35. The diagrams below represent fractured samples of four minerals.

Which mineral property is best illustrated by the samples?
(1) hardness   (2) streak   (3) cleavage   (4) density

36. Which New York State landscape region is composed mainly of metamorphosed surface bedrock?
(1) Taconic Mountains   (3) Atlantic Coastal Plain
(2) Allegheny Plateau   (4) Erie-Ontario Lowlands

Base your answers to questions 37 through 39 on the geologic cross section shown below.

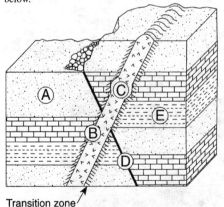

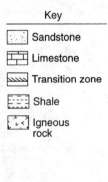

Key

Sandstone

Limestone

Transition zone

Shale

Igneous rock

Transition zone

37. At which location is metamorphic rock most likely to be found?
(1) A   (2) B   (3) C   (4) D

38. The most recently formed rock unit is at location
(1) A   (2) E   (3) C   (4) D

39. The graph below represents the percentage of each mineral found in a sample of rock *C*. Which mineral is in most likely represented by the letter *X* in the graph?

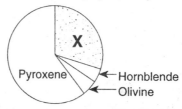

(1) potassium feldspar   (2) plagioclase feldspar   (3) quartz   (4) biotite

40. The photograph below represents a mountainous area in the Pacific Northwest.

Scientists believe that sedimentary rocks like these represent evidence of crustal change because these rocks were
(1) formed by igneous intrusion
(2) faulted during deposition
(3) originally deposited in horizontal layers
(4) changed from metamorphic rocks

41. How far from an earthquake epicenter is a city where the difference between the *P*-wave and *S*-wave arrival times is 6 minutes and 20 seconds?
(1) 1.7×10^3 km   (2) 9.9×10^3 km   (3) 3.5×10^3 km   (4) 4.7×10^3 km

42. The photograph below shows a large crater located in the southwestern United States.

Some fragments taken from the site have a nickel-iron composition. This evidence indicates that the crater probably was formed by
(1) the impact of a meteorite from space
(2) the collapse of a cavern roof
(3) an eruption of a volcano
(4) an underwater explosion of steam

43. Which map best represents the general pattern of magnetism in the oceanic bedrock near the mid-Atlantic Ridge?

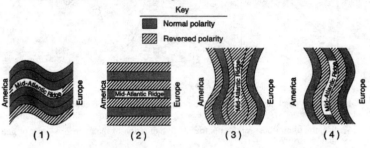

44. Which features are commonly formed at the plate boundaries where continental crust converges with oceanic crust?
    (1) large volcanic mountain ranges parallel to the coast at the center of the continents
    (2) a deep ocean trench and a continental volcanic mountain range near the coast
    (3) an underwater volcanic mountain range and rift valley on the ocean ridge near the coast
    (4) long chains of mid-ocean volcanic islands perpendicular to the coast

Base your answers to questions 45 and 46 on the diagrams below. Diagram I shows part of a geologic map. Diagram II shows a geologic cross section taken along line *CD*. The rock layers shown have not been overturned. Numbers 1 through 5 represent locations on the surface bedrock.

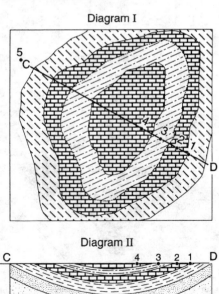

45. Which graph best represents the age of the surface bedrock along line *CD*?

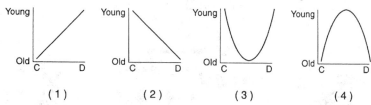

(1)   (2)   (3)   (4)

46. Which type of surface bedrock would most likely be found at location 5?
    (1) shale   (2) sandstone   (3) chemical limestone   (4) siltstone

47. Which column best represents the relative duration of the major intervals of geologic history?

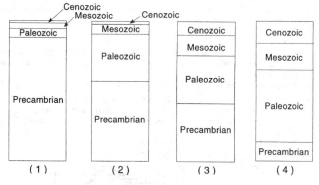

(1)   (2)   (3)   (4)

48. The diagram below represents a landscape area.

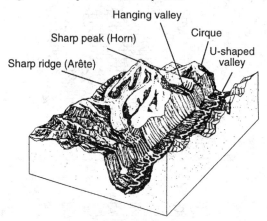

The labeled surface features of this landscape area resulted mainly from
(1) wind erosion   (2) wave erosion   (3) stream erosion   (4) glacial erosion

Base your answers to questions 49 through 51 on the map below, which shows an area of the northwestern United States affected by a major volcanic eruption at Crater Lake during the Holocene Epoch.

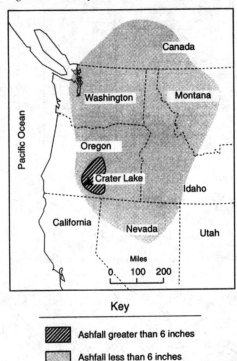

Key

Ashfall greater than 6 inches

Ashfall less than 6 inches

49. The pattern of distribution of the ash from the volcano was most likely caused by the direction of the
   (1) magnetic field
   (2) force of the volcanic eruption
   (3) flow of surface water
   (4) atmospheric air movements

50. The age of this volcanic eruption was most accurately determined to be Holocene by measuring the radioactive
   (1) potassium in the fine-grained volcanic rock
   (2) carbon in trees buried by the ash
   (3) uranium in the volcanic ash
   (4) rubidium in the igneous glass

51. This volcanic eruption is most useful to scientists today as a relative time marker in the geologic record of this map region because the
   (1) lava cooled quickly at the surface
   (2) lava contained radioactive rubidium-87
   (3) volcanic ash spread quickly over a large area
   (4) volcanic ash fell to Earth more quickly near the volcano than far from the volcano

52. The diagrams below represent the rock layers and fossils found at four widely separated rock outcrops.

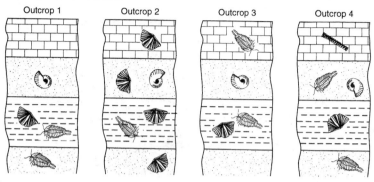

Which fossil appears to be the best index fossil?

(1)   (2)   (3)   (4)

53. The diagram below represents a landscape region and its underlying bedrock structure.

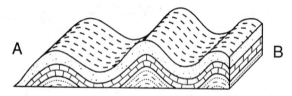

Which stream pattern is most likely present in this area?

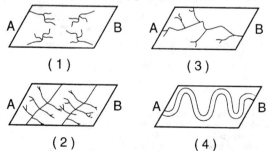

54. Which New York State landscape region is located at 42° N, 75° W?
    (1) Erie-Ontario Lowlands   (3) the Catskills
    (2) Hudson-Mohawk Lowlands   (4) Tug Hill Plateau

55. Which geologic processes produced the present surface landscape features of most New York State landscapes?
    (1) crustal movement and erosion   (3) faulting and folding
    (2) subsidence and metamorphism   (4) volcanism and igneous activity

PART II

This part consists of ten groups, each containing five questions. Choose seven of these ten groups. Be sure that you answer all five questions in each group chosen. [35]

Group 1

If you choose this group, be sure to answer questions 56–60.

Base your answers to questions 56 through 60 on the *Earth Science Reference Tables,* the diagrams below, and your knowledge of Earth science. The diagrams represent particles of the same type of sedimentary rock material collected from a streambed. The diagrams are drawn actual size.

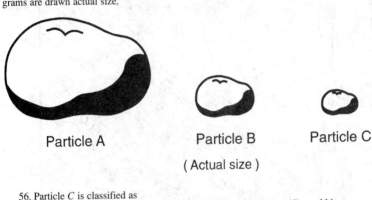

Particle A   Particle B   Particle C

(Actual size)

56. Particle *C* is classified as
    (1) sand   (2) a cobble   (3) a boulder   (4) a pebble

57. A student finds the mass of particle *B* to be 2.8 grams. The actual mass of the particle is 2.6 grams. What is the student's percent deviation (percent error)?
    (1) 7.1%   (2) 2.0%   (3) 7.7%   (4) 8.0%

58. Equal masses of each of the three particles are placed in a container of weak acid and shaken. Which particle size will weather most rapidly?
    (1) *A*   (3) *C*
    (2) *B*   (4) All samples will weather at the same rate.

59. Which inference about the density of particle *A* and particle *B* is most accurate?
    (1) Particle *A* and particle *B* have the same density because they are made of the same material.
    (2) Particle *A* has a greater density than particle *B* because particle *A* has a greater volume.
    (3) Particle *A* has a greater density than particle *B* because particle *A* has a greater mass.
    (4) Particle *B* has a greater density than particle *A* because particle *B* has been worn to a smaller size.

60. Particle *A* has a density of 2.7 grams per cubic centimeter and a volume of 15.0 cubic centimeters. What is the mass of this particle?
    (1) 5.5 g   (2) 15.0 g   (3) 40.5 g   (4) 109.3 g

Group 2

If you choose this group, be sure to answer questions 61–65.

Base your answers to questions 61 through 65 on the *Earth Science Reference Tables,* the contour map below, and your knowledge of Earth science. Points *A, B, C, D, X,* and *Y* are locations on the map. Elevations are expressed in feet. The maximum elevation of Basket Dome is indicated at point *X.*

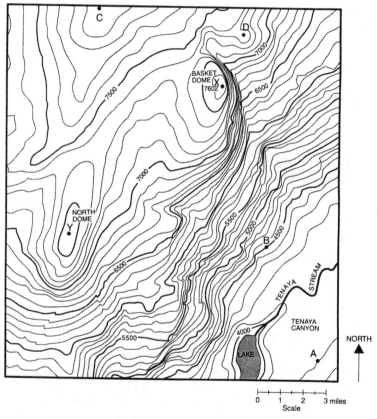

61. In which general direction does Tenaya Stream flow?
    (1) southeast to northwest
    (2) northwest to southeast
    (3) southwest to northeast
    (4) northeast to southwest

62. Which graph best represents the profile along a line between point B and point A?

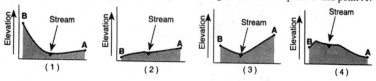

63. The highest elevation on the map is at point
    (1) *X*   (2) *Y*   (3) *C*   (4) *D*

64. The highest elevation of Basket Dome 40 years ago was measured at 7,600 feet. What is the rate of change in elevation for this area?
    (1) 0.6 in/yr   (2) 1.7 in/yr   (3) 24 in/yr   (4) 40 in/yr

65. Fossils of trilobites and eurypterids found in the rock near the top of Basket Dome provide evidence that this map area has most likely undergone
    (1) metamorphism from crustal plate collision
    (2) uplift from crustal plate movement
    (3) recent flooding from changes in worldwide sea level
    (4) volcanism from seafloor spreading

Group 3

If you choose this group, be sure to answer questions 66–70.

Base your answers to questions 66 through 70 on the *Earth Science Reference Tables,* the diagrams below, and your knowledge of Earth science. Diagram 1 shows a house located in New York State. Diagram II shows a solar collector that the homeowner is using to help heat the house.

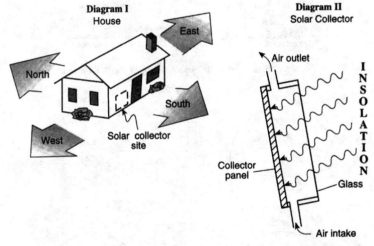

66. Air leaves the outlet of the solar collector because the air within the solar collector becomes
    (1) cooler and less dense   (3) warmer and less dense
    (2) cooler and more dense   (4) warmer and more dense

67. The homeowner decides to install carpet on the floor in the room that receives the most sunlight. A carpet with which characteristics would absorb the most insolation?
    (1) smooth texture and light color   (3) rough texture and light color
    (2) smooth texture and dark color   (4) rough texture and dark color

68. Which sequence best describes the pattern of energy transfer affecting the solar collector?
    (1) Sun (radiation) → collector panel (conduction and radiation) → air in collector (convection)
    (2) Sun (convection) → collector panel (convection and radiation) → air in collector (radiation)
    (3) Sun (conduction) → collector panel (conduction and convection) → air in collector (conduction)
    (4) Sun (conduction and convection) → collector panel (radiation) → air in collector (radiation)

69. For the angle of the Sun's rays shown, which side view best represents the correct placement of the solar collector to absorb the maximum amount of insolation?

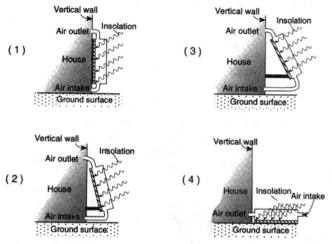

70. Which diagram best represents the apparent path of the Sun on June 21 for this location?

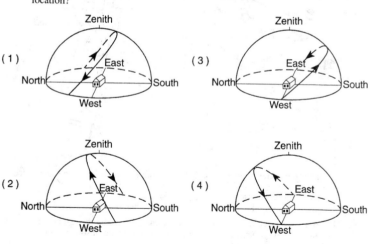

Group 4

If you choose this group, be sure to answer questions 71–75.

Base your answers to questions 71 through 75 on the *Earth Science Reference Tables*, the weather map below, and your knowledge of Earth science. The map shows a weather system that is affecting part of the United States.

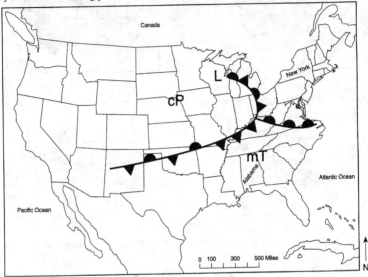

71. What is the total number of different kinds of weather fronts shown on this weather map?
    (1) 1   (2) 2   (3) 3   (4) 4

72. Compared to the air over most of the map region, the air mass centered over Alabama is
    (1) warmer and more humid   (3) colder and more humid
    (20 warmer and drier   (4) colder and drier

73. Which sequence of events forms the clouds associated with this weather system?
    (1) Moist air rises and becomes saturated in clean air.
    (2) Moist air rises, becomes saturated, and condenses on microscopic particles.
    (3) Moist air falls and reaches the dewpoint in clean air.
    (4) Moist air falls, reaches the dewpoint, and condenses on microscopic particles.

74. Which map best shows the areas in which precipitation is most likely occurring? [Darkened areas represent precipitation.]

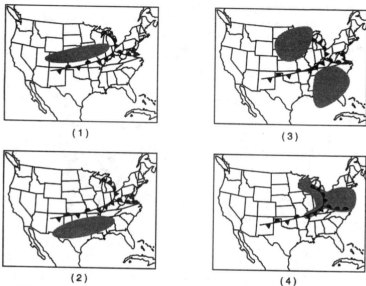

(1)

(3)

(2)

(4)

75. Which diagram shows the surface air movements most likely associated with the fronts?

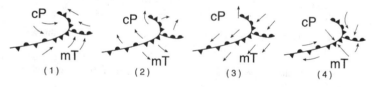

(1)

(2)

(3)

(4)

Group 5

If you choose this group, be sure to answer questions 76–80.

Base your answers to questions 76 through 80 on the *Earth Science Reference Tables,* the diagrams and descriptions of the two laboratory activities below, and your knowledge of Earth science. The particles used in these activities are described below.

Particles Used in Activities

Particle	Diameter	Density
	15 mm Al (aluminum)	2.7 g/cm^3
	10 mm Al (aluminum)	2.7 g/cm^3
	5 mm Al (aluminum)	2.7 g/cm^3

Particle	Diameter	Density
	15 mm Fe (iron)	7.9 g/cm^3
	15 mm Pb (lead)	11.4 g/cm^3

Activity 1

Three aluminum particles of different sizes were released in a plastic tube filled with water. The length of time each particle took to drop from point A to point B is shown in data table 1.

Data Table 1

Particle Size	Time of Settling
15 mm Al	3.2 sec
10 mm Al	5.4 sec
5 mm Al	7.2 sec

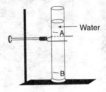

Activity 2

Different combinations of particles were placed in a tube filled with a thick liquid and allowed to fall to the bottom. The tube was then stoppered and quickly turned upside down, allowing the particles to settle. The different combinations of particles are shown in data table 2. The diagram of the particle sorting in data table 2 has been omitted intentionally.

Data Table 2

Combination	Particles Mixed	Diagram of Sorting
A	15 mm Al 10 mm Al 5 mm Al	
B	15 mm Al 15 mm Fe 15 mm Pb	

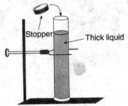

Note that question 76 has only three choices.

76. During Activity 1, as the 10-millimeter aluminum particle drops from A to B, the potential energy of the particle
   (1) decreases   (2) increases   (3) remains the same

77. In Activity 1, the three sizes of aluminum particles are placed in a stream with a velocity that will carry them all to a lake. Which cross section shows how the three sizes of particles are sorted when the stream slows as it empties into the lake?

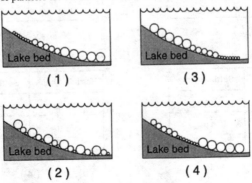

78. In Activity 2, when the tube is turned upside down, the particles of three different metals, labeled "Combination A," are allowed to settle. Which diagram represents the sorting that is most likely to occur?

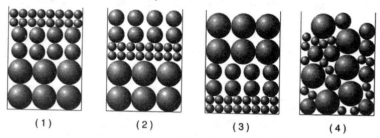

(1)   (2)   (3)   (4)

79. In Activity 2, when the tube is turned upside down, the particles of three different metals, labeled "Combination B," are allowed to settle. Which diagram represents the sorting that is most likely to occur?

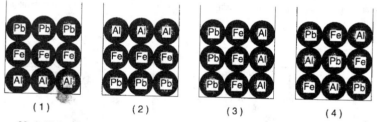

(1)   (2)   (3)   (4)

80. A third activity, similar in setup to Activity 1, was done using flat, oval, and round aluminum particles with identical masses. Which table shows the most likely results of this third activity?

Particle Shape	Settling Time
Round	5.1 sec
Oval	5.1 sec
Flat	5.1 sec

(1)

Particle Shape	Settling Time
Round	6.7 sec
Oval	5.1 sec
Flat	3.2 sec

(3)

Particle Shape	Settling Time
Round	5.1 sec
Oval	3.2 sec
Flat	6.7 sec

(2)

Particle Shape	Settling Time
Round	3.2 sec
Oval	5.1 sec
Flat	6.7 sec

(4)

Group 6

If you choose this group, be sure to answer questions 81–85.

Base your answers to questions 81 through 85 on the *Earth Science Reference Tables*, the diagram below, and your knowledge of Earth science. The diagram shows the structure of a student-developed chart for identifying some rock samples. The circles labeled choice 1 through choice 4 represent decisionmaking steps leading either to path (*a*) or path (*b*). Choice 5 has not been completed.

Student Chart

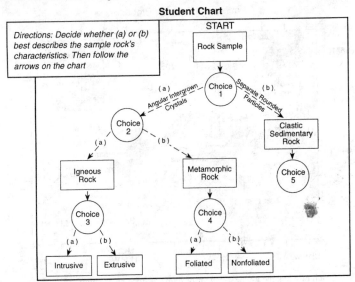

81. Before the student can select either path (*a*) or path (*b*) at choice 1, the student must make a decision about
    (1) mineral composition
    (2) crystal size
    (3) the temperature at which rocks form
    (4) the appearance of the rock grains

82. At choice 2, the student should generally select path (*a*) if the student observes
    (1) a random arrangement of mineral crystals
    (2) distorted structure and crystal alignment
    (3) bands of mineral crystals
    (4) layers of same-sized crystals

83. Which rock specimen men should lead the student to choice 4, path (*a*)?
    (1) peridotite   (2) quartzite   (3) gneiss   (4) dolostone

84. Which characteristic should be used at choice 5 to further identify the types of clastic sedimentary rocks?
    (1) grain size
    (2) mineral cement
    (3) mineral color
    (4) horizontal layering

85. The chart is best described as
    (1) a rock cycle diagram
    (2) a classification system
    (3) an erosional-depositional system
    (4) a mineral identification diagram

Group 7

If you choose this group, be sure to answer questions 86–90.

Base your answers to questions 86 through 90 on the *Earth Science Reference Tables,* the map below, and your knowledge of Earth science. The map shows a portion of California along the San Andreas Fault zone. The map shows the probability (percentage chance) that an earthquake strong enough to damage buildings and other structures will occur between now and the year 2024.

Earthquake Damage Probability

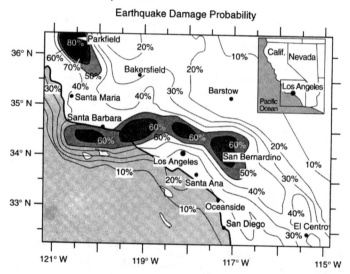

86. Which city has the greatest danger of damage from an earthquake?
    (1) Barstow   (2) Parkfield   (3) Oceanside   (4) San Bernardino

87. This fault zone is located along the boundary between which two crustal plates?
    (1) Cocos plate and Pacific plate
    (2) North American plate and Pacific plate
    (3) Nazca plate and Cocos plate
    (4) North American plate and South American plate

88. If a large earthquake were to occur at San Diego, the earliest indication at another California location of the occurrence of that earthquake would be the arrival of the
    (1) S-waves at Oceanside   (3) P-waves at Oceanside
    (2) S-waves at San Bernardino   (4) P-waves at San Bernardino

89. Which diagram best represents the relative movements of the crustal plates along the San Andreas Fault in the map area?

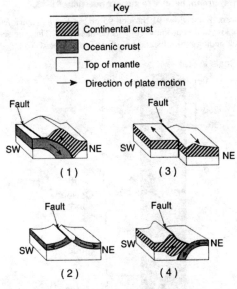

90. Which map best represents the location of the primary San Andreas Fault line?

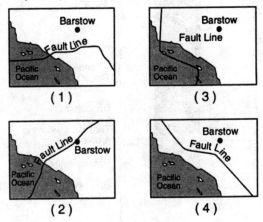

Group 8

If you choose this group, be sure to answer questions 91–95.

Base your answers to questions 91 through 95 on the *Earth Science Reference Tables,* the core section below, and your knowledge of Earth science. The core section shows the subsurface bedrock geology for a location north of Buffalo, New York.

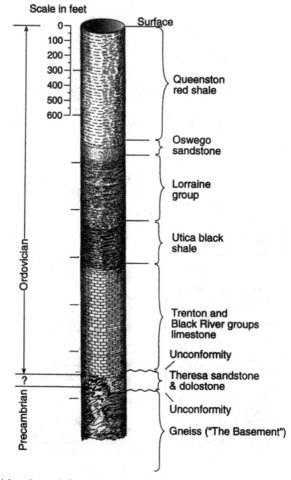

91. Which rock type is found at a depth of 1,800 feet at this location?
    (1) Oswego sandstone   (3) Trenton limestone
    (2) "Basement" gneiss   (4) Theresa sandstone

92. What do the unconformities shown near the base of the drill core indicate?
    (1) The continental plates were separated for a long period of time.
    (2) Part of the geologic rock record has been destroyed.
    (3) This area was covered by a warm, shallow sea.
    (4) Extinction of many kinds of living things was widespread.

93. Which "Important Fossils of New York" are most likely to be found in the Utica black shale or the Black River limestone?

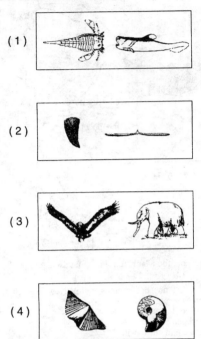

94. Which statement best explains why more fossils are found in outcrops of Black River rocks than in outcrops of Utica shales?
    (1) Life-forms lacked hard parts at the time of Black River deposition.
    (2) Many fossils of the Utica shales were destroyed by metamorphism.
    (3) The Black River group was deposited in an environment that supported more life-forms.
    (4) The Utica shales were deposited over a wider geographic area.

95. Based on studies of fossils found in the Trenton group, scientists have estimated that the climate of New York State during this part of the Ordovician Period was much warmer than the present climate. Which statement best explains this change in climate?
    (1) The North American Continent was nearer to the Equator during the Ordovician Period.
    (2) The Sun emitted less sunlight during the Ordovician Period.
    (3) Earth was farther from the Sun during the Ordovician Period.
    (4) Many huge volcanic eruptions occurred during the Ordovician Period.

Group 9

If you choose this group, be sure to answer questions 96–100.

Base your answers to questions 96 through 100 on the *Earth Science Reference Tables,* the map below, and your knowledge of Earth science. The map shows the Generalized Landscape Regions of New York State as they appear in the Earth Science Reference Tables. Letters *A* through *K* represent the different landscape regions. Letters *P* and *Q* indicate viewpoints for interpreting landscape cross sections. Letters *X* and *Y* are two points along the New York–New Jersey border.

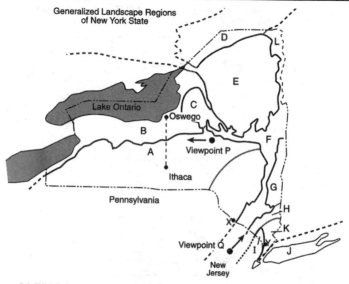

Generalized Landscape Regions
of New York State

96. Which letter represents the Manhattan Prong landscape region?
(1) *H*　　　　(2) *I*　　　　(3) *J*　　　　(4) *K*

97. The location of these landscape regions within New York State is mostly determined by differences in regional
(1) human population densities
(2) climate characteristics
(3) bedrock structure and composition
(4) rock age and stream-drainage patterns

98. Which features within these landscape regions have developed primarily because the underlying rock is resistant to weathering and erosion?
(1) lowlands and plains　　　　(3) valleys and hilltops
(2) hilltops and plains　　　　(4) escarpments and highlands

99. As seen from viewpoint *P*, which bedrock cross section best illustrates the geologic changes that occur between Ithaca in landscape region *A* and Oswego in landscape region *B*? [Cross sections are not drawn to scale.]

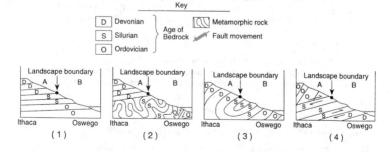

100. Which cross section best illustrates the surface landscape and the underlying bedrock across line *XY* as seen from viewpoint Q?

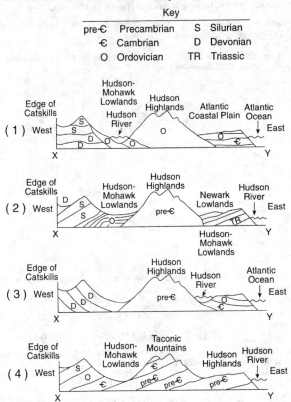

Key

pre-Є	Precambrian	S	Silurian
Є	Cambrian	D	Devonian
O	Ordovician	TR	Triassic

Group 10

If you choose this group, be sure to answer questions 101–105.

Base your answers to questions 101 through 105 on the *Earth Science Reference Tables* and on your knowledge of Earth science.

101. Which model of a planet's orbit best represents the actual eccentricity of the orbit of Mars? [Models are drawn to scale.]

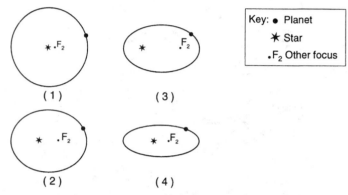

102. What is the total number of calories required to melt 100 grams of ice at 0°C to liquid water at 0°C?

(1) 5,400 cal   (2) 8,000 cal   (3) 54,000 cal   (4) 80,000 cal

103. Which member of the solar system has an equatorial diameter of 3.48×10^3 kilometers?

(1) Moon   (2) Earth   (3) Sun   (4) Pluto

104. What is the approximate age of an igneous rock that contains only one-fourth of its original potassium-40 content due to radioactive decay?

(1) 1.3×10^9 years   (3) 3.9×10^9 years
(2) 2.6×10^9 years   (4) 5.2×10^9 years

105. The diagram below shows a cross section of some Devonian-age rocks along the western side of the mid-Hudson River Valley.

Which two mountain-building episodes could have been responsible for deforming these rock layers?

(1) Grenville and Taconian orogenies
(2) Taconian and Acadian orogenies
(3) Acadian and Appalachian orogenies
(4) Appalachian and Grenville orogenies

(Continued from other side)

5. HUDSON-MOHAWK LOWLANDS. The rocks of this region are distorted sedimentary rocks of Paleozoic age that have lower resistance than the surrounding rocks. They have therefore been leveled by weathering, and by the erosion of the Mohawk and Hudson Rivers, to generally low elevations. Variations in slope are associated with escarpments that have been formed by resistant layers and the valleys carved by the rivers. South of Albany the Hudson River has an elevation of sea level and may be considered an inlet of the ocean.

6. ALLEGHENY PLATEAU. This part of the Appalachian Plateau is composed of horizontal sedimentary rocks, mostly of Paleozoic age. Much leveling by streams and glaciers has resulted in steep slopes and much change in elevation, so that some sections, such as the Catskills (area 6B), resemble mountains. Elevations range from about 500 meters in the west to over 1,250 meters in the Catskills. The region contains many lakes, such as the Finger Lakes, which are the result of glacial erosion and deposition.

7. ERIE-ONTARIO LOWLANDS. This region is a plain of horizontal sedimentary rocks of Paleozoic age, covered by much glacial transported sediment. The northern section is especially smooth because of deposition of sediments from glacial lakes that were the ancestors of Lakes Erie and Ontario. In some places, resistant rocks have been eroded to escarpments up to 500 meters in elevation, but elevations and slopes are generally small throughout the region.

8. TUG HILL PLATEAU. This plateau of horizontal sedimentary rocks of Paleozoic age has a resistant surface layer that has resulted in fairly uniform elevations of about 600 to 700 meters. However, steep slopes occur where rivers have cut valleys, such as the Black River Valley that separates this region from the Adirondacks.

9. ADIRONDACK MOUNTAINS. These are mature mountains composed mostly of metamorphic rocks of Precambrian age. There are moderate elevations in the western section, but the highest elevations in the state (over 1,600 meters at mountain peaks) occur in the eastern section. The landscape is rugged, with much change in elevation and steep slopes, and with valleys related to faults and rocks of lesser resistance. Stream drainage patterns are of radial and trellis types except where glacial deposition has blocked the former drainage paths, resulting in many swamps and lakes.

10. ST. LAWRENCE-CHAMPLAIN LOWLANDS. This plain is composed generally of horizontal sedimentary rocks of lower resistance than rocks of surrounding regions, and therefore has lower elevations. The St. Lawrence section (10 A) is mostly flat, with changes in elevation usually no more than 30 meters. Some steep slopes exist in the Champlain section (10 B) as the result of uplift and subsidence caused by faults. Lake Champlain is the site of a subsided block.

LANDSCAPE REGIONS OF NEW YORK STATE AND THEIR CHARACTERISTICS

New York has a greater number of different landscape regions than any other state. The chief reason for this variety of landscape regions is the great variation in age, structure, and resistance of the bedrock found within the state. The present climate of the state is uniformly humid and has not been a major factor in producing variations in landscape development.

Almost all of New York was affected by the Pleistocene glaciation that ended in New York about 10,000 years ago. As a result, glacial depositional features are observed throughout the state, superimposed on the other characteristics of the different landscape regions. One of the most important features left by the glaciers is the transported soils, which are young and have small and incomplete profiles.

The landscape regions that are numbered on the facing map have the following distinguishing characteristics.

1. ATLANTIC COASTAL PLAIN. The bedrock consists of horizontal sedimentary rocks of late Mesozoic and Cenozoic age, with elevations near sea level, covered by thick glacial deposits reaching elevations of about 125 meters above sea level in some areas. The features of glacial deposition result in minor changes of slope on a generally smooth plain. Waves and ocean currents have created many typical shore features, such as bars, beaches, and lagoons.

2. NEWARK LOWLANDS. This region is composed of weak sedimentary rocks of early Mesozoic age, which have been leveled to lower elevations than the surrounding landscape regions. The generally smooth landscape has moderate changes in slope caused by faults. A volcanic intrusion called the Palisades Sill borders the Hudson River. Because of its greater resistance, it forms a cliff that ranges up to more than 150 meters above sea level.

3. HUDSON HIGHLANDS—MANHATTAN PRONG. These two regions are extensions of the New England Highlands to the east. These mountains in the mature and old-age stages are composed of highly distorted metamorphic rocks, mostly early Paleozoic and Precambrian in age and more resistant than the rocks of surrounding regions. In the northern parts of the region (3A, Hudson Highlands) there are elevations up to 500 meters with steep slopes; in the southern parts (3B, Manhattan Prong), the rocks have been eroded to low elevations and little slope typical of the plain-like topography of mountains in old age.

4. TACONIC MOUNTAINS. These greatly eroded mountains were originally uplifted in the Paleozoic Era. They consist mostly of metamorphic nonsedimentary rocks that are highly folded and faulted. Today they have moderate elevations (up to about 600 meters) and gradual changes in slope, so that the topography has the form of rolling hills.

(Continued on other side)